# A Discrete Event Approach for Model-Based Location Tracking of Inhabitants in Smart Homes

Ein ereignisdiskreter Ansatz zur modellbasierten Lokalisierung
der Bewohner in intelligenten Wohnungen

Une approche orientée événements discrets pour la localisation des habitants
dans des habitats intelligents basée sur le modèle

Vom Fachbereich Elektrotechnik und Informationstechnik
der Technischen Universität Kaiserslautern
zur Verleihung des akademischen Grades

**Doktor der Ingenieurwissenschaften (Dr.-Ing.)**

genehmigte Dissertation

von
**Mickaël Danancher**
geb. in Viriat (Frankreich)

D 386

Eingereicht am: 01.10.2013
Tag der mündlichen Prüfung: 02.12.2013
Dekan des Fachbereichs: Prof. Dr.-Ing. Hans D. Schotten
(Technische Universität Kaiserslautern)

Promotionskommission:
Vorsitzender: Prof. Dr. habil. Etienne Craye
(Ecole Centrale de Lille, Frankreich)
Prof. Dr.-Ing. habil. Lothar Litz
(Technische Universität Kaiserslautern)

Berichterstattende: Prof. Dr. habil. Jean-Jacques Lesage
(Ecole Normale Supérieure de Cachan, Frankreich)
Prof. Dr.-Ing. habil. Lothar Litz
(Technische Universität Kaiserslautern)
Prof. Dr. Maria Pia Fanti
(Politecnico di Bari, Italien)
Prof. Dr.-Ing. Jan Lunze
(Ruhr-Universität Bochum)

Bibliographic information published by the Deutsche Nationalbibliothek

The Deutsche Nationalbibliothek lists this publication in the Deutsche
Nationalbibliografie; detailed bibliographic data are available
in the Internet at http://dnb.d-nb.de .

ISBN 978-3-8325-3634-3

Logos Verlag Berlin GmbH
Comeniushof, Gubener Str. 47,
10243 Berlin
Tel.: +49 (0)30 42 85 10 90
Fax: +49 (0)30 42 85 10 92
INTERNET: http://www.logos-verlag.de

**Abstract** - Life expectancy has continuously increased in most industrialized countries over the last decades and will probably continue to increase in the future. This leads to new challenges relative to the autonomy and the independence of elderly. The development of Smart Homes is a direction to face these challenges and to enable people to live longer in a safe and comfortable environment. Making a home smart consists in placing sensors, actuators and a controller in the house in order to take into account the behavior of their inhabitants and to act on their environment to improve their safety, health and comfort. Most of these approaches are based on the real-time indoor Location Tracking of the inhabitants. In this thesis, a whole new approach for model-based Location Tracking of an *a priori* unknown number of inhabitants is proposed. This approach is based on Discrete Event Systems paradigms, theory and tools. The usage of Finite Automata (FA) to model the detectable motion of the inhabitants as well as different methods to create such FA models have been developed. Based on these models, algorithms to perform efficient Location Tracking are defined. Finally, several approaches aiming at evaluating the relevance of the instrumentation of a Smart Home with the objective of Location Tracking are proposed. Throughout the thesis, the different contributions are illustrated on case studies. The approach has also been fully implemented and tested.

**Résumé** - L'espérance de vie a augmenté dans les dernières décennies et devrait continuer à croître dans les prochaines années. Cette augmentation entraîne de nouveaux défis concernant l'autonomie et l'indépendance des personnes âgées. Le développement d'habitats intelligents est une piste pour répondre à ces défis et permettre aux personnes de vivre plus longtemps dans un environnement sûr et confortable. Rendre un habitat intelligent consiste à y installer des capteurs, des actionneurs et un contrôleur afin de pouvoir prendre en compte le comportement de ses habitants et agir sur leur environnement, pour améliorer leur sécurité, leur santé et leur confort. La plupart de ces approches s'appuient sur la localisation en temps réel des habitants dans leur habitat. Dans cette thèse, une nouvelle approche complète permettant la localisation d'un nombre *a priori* inconnu d'habitants basée sur le modèle est proposée. Cette approche tire parti des paradigmes, de la théorie et des outils des Systèmes à Événements Discrets. L'utilisation des automates à états finis pour modéliser le mouvement détectable des habitants ainsi que des méthodes permettant de construire ces modèles ont été développées. A partir de ces modèles automates finis, plusieurs algorithmes permettant de localiser de manière efficace les habitants ont été définis. Enfin, plusieurs approches pour l'évaluation des performances de l'instrumentation d'un habitat intelligent pour un objectif de localisation ont été proposées. Tout au long de cette thèse, les différentes contributions sont illustrées à l'aide de cas d'étude. La méthode a également été totalement implémentée et mise à l'épreuve.

**Zusammenfassung** - In den meisten Industrieländern ist die Lebenserwartung in den letzten Jahrzehnten fortlaufend gestiegen und wird höchstwahrscheinlich noch weiter steigen. Dieser Anstieg führt zu neuen Herausforderungen hinsichtlich der Autonomie und Unabhängigkeit von älteren Menschen. Die Entwicklung von intelligenten Wohnungen ist ein Weg diesen Herausforderungen zu begegnen und es den Menschen zu ermöglichen länger in einer sicheren und komfortablen Umgebung zu leben. Dazu stattet man solche Wohnungen mit Sensoren, Aktoren sowie einem Controller aus. Dies ermöglicht es, in Abhängigkeit vom Verhalten der Bewohner, die Wohnumgebung so zu beeinflussen, dass sich Sicherheit, Gesundheit und Komfort verbessern. Ansätze, die dies zum Ziel haben, basieren meistens auf Methoden, die es ermöglichen Menschen innerhalb ihrer Wohnung in Echtzeit zu lokalisieren. In dieser Dissertation wird daher ein neuer Ansatz für eine modellbasierte Lokalisierung einer *a priori* unbekannten Anzahl von Bewohnern vorgestellt. Dieser Ansatz fußt auf der Theorie, den Paradigmen und den Werkzeugen aus dem Gebiet der ereignisdiskreten Systeme. Es werden endliche Automaten eingesetzt, um die von den Sensoren erfassbaren Bewohnerbewegungen zu modellieren. Verschiedene Verfahren zur Erzeugung solcher Automaten werden gezeigt. Basierend auf diesen Modellen werden Algorithmen definiert, mittels derer die Bewohner wirksam lokalisiert werden können. Abschließend werden Methoden vorgeschlagen, die dazu dienen die Relevanz der Sensorinstrumentierung für die Lokalisierung zu bewerten. Die entwickelten Verfahren werden in der Dissertation durchgehend anhand von Fallbeispielen erläutert. Der gesamte Ansatz wurde implementiert und erprobt.

*"Big Brother is watching you"*

Nineteen Eighty-Four, George Orwell, 1949

# Acknowledgments

Even if a Ph.D. thesis is a personal work, all of this would not be possible without the help of several people. May they find here the expression of my deepest gratitude.

First of all, I want to thank my two doctoral advisors Professor JEAN-JACQUES LESAGE and Professor LOTHAR LITZ. They gave me the intellectual freedom I needed and they were always there to discuss new ideas. Having two advisors also brought me two different perspectives on our research field and their complementary points of view were a real chance for me. I enjoyed working with them during all these years, either in Cachan or in Kaiserslautern.

I would also like to thank the two external reviewers of my Ph.D. thesis, Professor MARIA PIA FANTI and Professor JAN LUNZE. It was a pleasure and an honor for me that they accepted to review this thesis.

I want to give many thanks to Professor ETIENNE CRAYE who accepted to participate in the doctoral committee and to be its president, despite the strong constraints on the date of the defense. I am also grateful to him for conducting in a pleasant way the discussion after my presentation.

I would also like to thank in particular one colleague: JEREMIE SAIVES, who was an attentive proofreader of this manuscript and noted numerous imprecise explanations and grammar mistakes. This was a precious help for me.

I am also grateful to the three students I advised during my thesis: MATTHIAS BORDRON, SIMON RADEL and JULES SCORDEL. The interest they showed for my work and their involvement in the development of the Smart Home Emulator helped me to get significant results.

As a Ph.D. student in cotutorship, I had colleagues in two institutes, the Lurpa in France and AT+ in Germany. I spent many good times with all of them, current and former colleagues. I will surely miss our numerous discussion about more or less scientific subjects. Special thanks for my French colleagues of the CIVIL team, always there for having fun, and for my German colleagues who made me discover the German way of life and always made me feel welcome when I spent time in Germany. I particularly want to thank MONIKA KUNZ, our highly valuable secretary in AT+, who helped me with the numerous formalities and with the organization of my stays in Germany. I would also like to give special thanks to the colleagues who shared the office with me, in Kaiserslautern and in Cachan, especially during the writing of this thesis.

My friends were also of primary importance. They supported me and allowed me to get some fresh air and detach myself from the science from time to time, either in Paris or elsewhere in France (Nantes, Haguenau, Antibes, La Rivaudière...).

And finally, of course, I want to express my deepest gratitude to my family: my parents, my two sisters and CACO. They were always behind me and always comforting me in my choices. Their presence during my Ph.D. defense meant a lot to me.

Cachan, December 2013

MICKAËL DANANCHER

# Contents

# List of Figures

# List of Tables

# List of Algorithms

# Frequently used abbreviations

| | |
|---|---|
| **AAL** | Ambient Assisted Living |
| **ADL** | Activity of Daily Living |
| **CM** | Confusion Matrix |
| **DES** | Discrete Event System |
| **DMA** | Detectable Motion Automaton |
| **FA** | Finite Automaton |
| **FDI** | Fault Detection and Isolation |
| **LT** | Location Tracking |
| $\mathbf{MIDMA}_N$ | Multiple ($N$) Inhabitants Detectable Motion Automaton |
| $\mathbf{MIDMA}_N^{red}$ | Reduced Multiple ($N$) Inhabitants Detectable Motion Automaton |
| **MILT** | Multiple Inhabitants Location Tracking |
| $N$**-ALA** | $N$ Accurate Location Ability |
| $N$**-ILT** | $N$ Inhabitants Location Tracking |
| $N_{max}$ | Maximal number of trackable inhabitants |
| **SILT** or **1-ILT** | Single Inhabitant Location Tracking |

# Introduction

## Motivation

Life expectancy has continuously increased in most industrialized countries over the last decades and will probably continue to increase in the future. Moreover, according to (Eurostat, 2010) and Fig. 0.1, the age structure of the population in the European Union is aimed to change drastically and will lead to a bigger proportion of elderly.

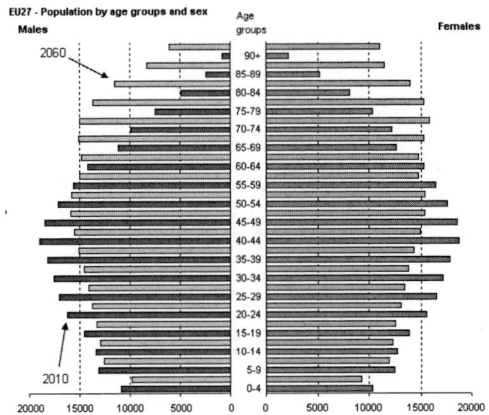

Figure 0.1.: Age structure of the population in 2010 and 2060 in the EU (Eurostat, 2010)

This situation is not limited to the European Union and according to (World Health Organization, 2012) and Fig. 0.2, the percentage of people aged 60 or over will reach 30 % in many countries in 2050.

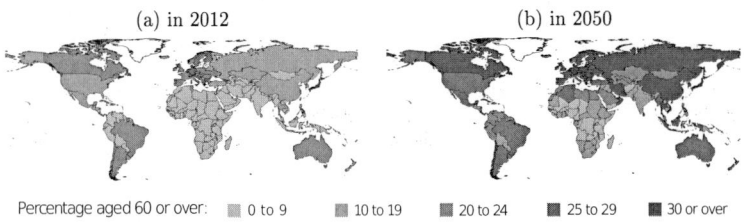

Figure 0.2.: Percentage of the population aged 60 or over in 2012 (a) and in 2050 (b) (World Health Organization, 2012)

These considerations about aging lead to new issues concerning the autonomy and the independence of elderly or disabled people. Ambient Assisted Living (AAL) technologies are aiming to help them to live autonomously in a comfortable and safe environment. These technologies are mostly based on the development of smart environments, for instance Smart Homes.

A possible representation of a Smart Home is given in Fig. 0.3. It can be seen as a closed loop between the inhabitants living in a home and a controller. The controller gets information about the behavior of the inhabitants through the sensors installed in the house and may modify the environment by acting on the actuators of the house. This representation is inspired by the representation of a closed-loop Discrete Event System (DES) where a controller is sensing and acting on a plant respectively through the sensors and the actuators.

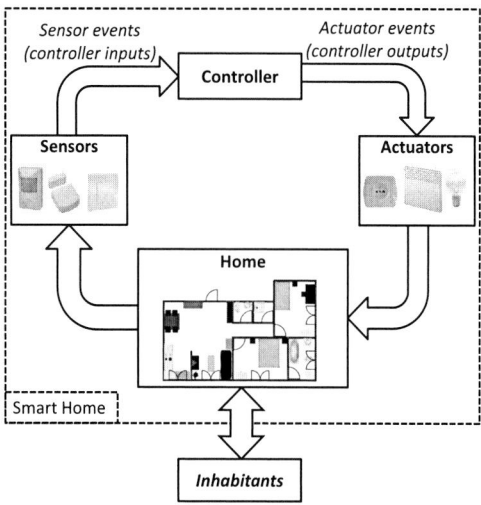

Figure 0.3.: Smart Home seen as a closed-loop Discrete Event System

Smart Homes are meant to improve the quality of life of the inhabitants and thus cover different possible applications. The most developed approaches are relative to the comfort of the inhabitants, their safety and health problem detection. Comfort can be improved in a Smart Home by performing for instance a smart monitoring of the energy consumption or of the entertainment of the inhabitants (music, TV) according to their habits or their life style. Safety applications cover different aspects, such as the detection of fire or water flooding, the detection of burglars or the automatic shutdown of dangerous devices when the inhabitant is leaving the house or sleeping. Finally health is improved through the more or less quick detection of health problems (fall, unconsciousness, and long term diseases like dementia) and the transmission of an alarm to caregivers or emergency services depending on the level of emergency.

Several of these approaches are requiring the location of the inhabitant(s) to be known at any time in order to perform well. For instance, an approach aiming at automatic shutdown of dangerous devices requires to know when a "potentially dangerous" room is empty i.e. when all the inhabitants are out of this room or outside the house. Regarding the health problem detection, some zones of the house may be more critical than other ones (the bathroom for instance) and thus, the location of one inhabitant in this particular room needs to be known. Estimating at any time the location of one or several inhabitants is named **Location Tracking**.

Existing Location Tracking approaches are functional, however they show some limits. Most of them are not based on a formal and explicit model because based on a learned model. By using such learning methods, the implicit model they provide is highly depending on the current installed sensors and on the current inhabitants living in the house. If one of these two parameters (sensors or inhabitants) changes, the model should be built again. As a drawback, a potentially long learning phase is required for the first modeling or each time a new model should be built, thus delaying the availability of the Location Tracking approach.

The objective of this thesis is to propose a new approach for single or multiple inhabitants indoor Location Tracking using explicit models not depending on the inhabitants' behavior, thus being robust to any inhabitants' behavior and not requiring a learning phase to be built. To achieve this goal, three contributions relative to **modeling**, **tracking** and **evaluating** are proposed. They are summarized in the next section and will be detailed along this manuscript.

## Contributions of the thesis

The contributions of the present thesis are summarized in Fig. 0.4.

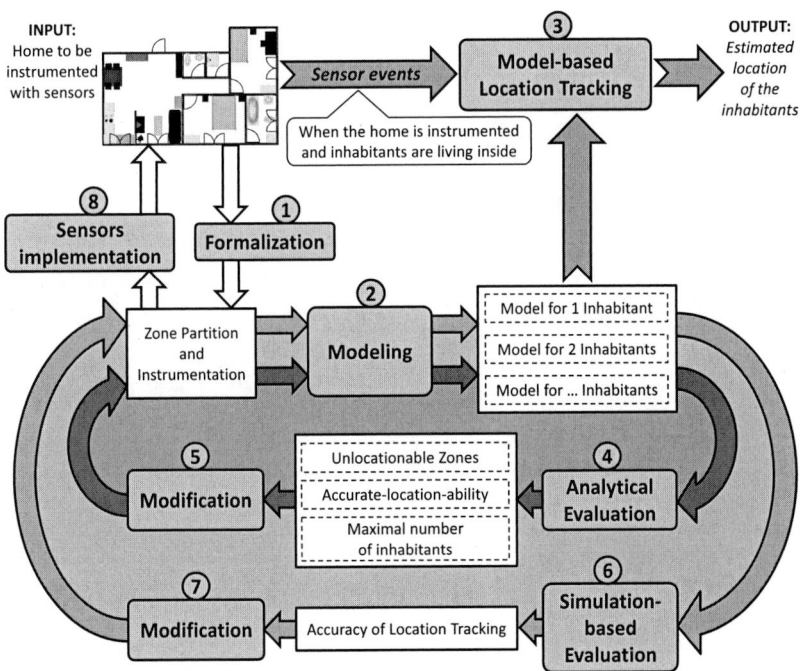

Figure 0.4.: Overview of the contributions of the thesis

In a first step, a model of the detectable motion of one inhabitant or of several inhabitants in a Smart Home is defined. Furthermore, different methods for building this model for a given Smart Home independently of the inhabitants living inside are proposed. This modeling step constitutes the first contribution and is represented by the steps 1 and 2 of Fig. 0.4.

In a second time, these models (model of the detectable motion for one inhabitant, model of the detectable motion for $N$ inhabitants) constitute the inputs for model-based Location Tracking. First, the particular cases of Single Inhabitant Location Tracking or of *a priori* known and fixed number of inhabitants Location Tracking are treated. Then they are extended to the Location Tracking of an *a priori* unknown number of inhabitants. Finally, the impact of potential sensor faults on the Location Tracking and the possibility of fault-tolerant Location Tracking are discussed. This Location Tracking step constitutes the second contribution of the thesis and is represented by the step 3 of Fig. 0.4.

Finally, the relevance of a given Smart Home for performing Location Tracking of a given maximal number of inhabitants has to be evaluated. Thus, an approach for analytical evaluation and another one for simulation-based evaluation of the performances is proposed. Based on these two approaches, a procedure for evaluation-aided improvement of Smart Home instrumentations is developed. The analytical evaluation approach constitutes the step 4 of Fig. 0.4, the simulation-based evaluation approach constitutes the step 6, and finally, the evaluation-aided improvement procedure is described by the loops 2, 4, 5 and 2, 6, 7 and by the feedback step 8.

## Organization of the manuscript

The manuscript is organized as follows. A review of the state of the art in the field of Ambient Assisted Living and Smart Home is given in Chapter 1. In this chapter, the concept of Smart Home, Ambient Assisted Living and Location Tracking are defined and several existing applications, methods and sensors are briefly described. This chapter ends with the problem statement of indoor online Location Tracking and a summary of the assumptions and considerations that will hold in this thesis.

In Chapter 2, the first contribution relative to the proposition of a detectable motion model for a single inhabitant and for multiple inhabitants is detailed. Several modeling procedures are also proposed and compared on a case study.

In Chapter 3, the different developed procedures to perform online Location Tracking (of an *a priori* known number of inhabitants or of an *a priori* unknown number of inhabitants) are described. An approach for sensor Fault Detection and Isolation is also given and the possibility of fault-tolerant Location Tracking is discussed.

In Chapter 4, the procedures and criteria to evaluate (analytically or based on simulation) the ability of an instrumentation to perform Location Tracking are detailed. Moreover, a procedure for evaluation-aided improvement of Smart Homes is given.

In the last chapter, a summary of the contributions and outlook for future work are given.

Finally, in the appendix of this thesis, the implementation of a Smart Home emulator and its usage for simulation-based performance evaluation is detailed.

# 1. Ambient Assisted Living, Smart Home and indoor Location Tracking

## Introduction

As stated in the introduction, technologies aiming to help people to live autonomously in a comfortable and safe home currently represent an emerging field of research and innovation. Depending on the needs of the inhabitants and of the societal context, these technologies may cover different applications concerning for instance the comfort, the safety or the health of the inhabitants. To achieve such applications, different approaches based on different scientific backgrounds and different considerations on the sensors that may be installed were developed.

This chapter is aiming to review the state of the art relative to this topic. In a first section, the concepts of Ambient Assisted Living (AAL) and Smart Home are detailed. Considerations about the acceptance of such technologies by the inhabitants are also presented. In a second section, an overview of the possible applications of such technologies is given. The different techniques developed in order to comply with these applications are presented in the third section and a review of the different sensors existing in the literature is given in the fourth section. Finally, the objective of the present thesis, its position regarding the other already existing approaches and the problem statement of explicit-model-based Location Tracking conclude this chapter.

## 1.1. Ambient Assisted Living and Smart Home

### 1.1.1. Definitions

The term AAL first appeared in the European Commission's terminology in the early 2000s (Floeck, 2010). Different definitions have since been proposed. For instance, according to the Ambient Assisted Living Joint Programme of the European Commission (AAL JP, 2013), the concept of Ambient Assisted Living is understood as:

- to extend the time people can live in their preferred environment by increasing their autonomy, self-confidence and mobility;

- to support maintaining health and functional capability of the elderly individuals;

- to promote a better and healthier lifestyle for individuals at risk;

- to enhance the security, to prevent social isolation and to support maintaining the multifunctional network around the individual;

- to support carers, families and care organizations;

- to increase the efficiency and productivity of used resources in the aging societies.

AAL is mainly devoted to the people in their old age and their relative (families or care givers). A simplified definition is the following:

**Definition 1** (Ambient Assisted Living AAL). Ambient Assisted Living is a concept enabling (senior) people to live independently in their accustomed homes and their social environments as long as possible by supporting their daily life with technical devices.

According to (Floeck, 2010), the AAL functionalities are divided into three main categories: Health, Safety and Comfort.

Assistance to elderly may also have other names in the literature, for instance, in (Kleinberger et al., 2007), the notion of Home Care System is defined as follows:

**Definition 2** (Home Care System HCS (Kleinberger et al., 2007)). The aim of a Home Care System is to allow the assisted persons to live longer in their preferred environment at home, while retaining their independence, even when they have handicaps or medical diseases.

This concept is close to the concept of AAL and the Fig. 1.1 confirms the same majors points which are emergency assistance (health and safety), autonomy enhancement (health) and comfort.

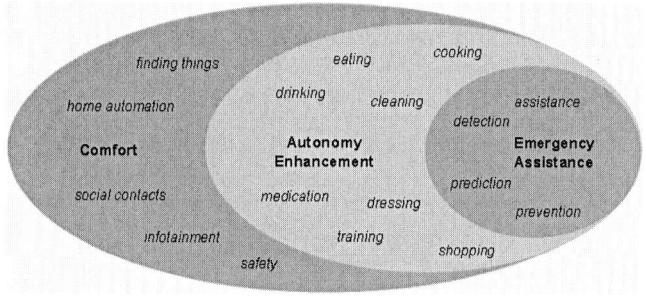

Figure 1.1.: Home Care System domain (Kleinberger et al., 2007)

A more generic term to cover this topic is "Smart Home". The concept of Smart Home is widely used and several definitions exist. A global one can be found in (Robles and Kim, 2010) where Smart Home is defined as follows:

**Definition 3** (Smart Home (Robles and Kim, 2010)). Smart Home is the integration of technology and services through home networking for a better quality of living.

Moreover, the authors indicate that *Smart Home is the term commonly used to define a residence that uses a Home Controller to integrate the residence's various home automation systems* (Robles and Kim, 2010).

In (Cook and Das, 2007), the authors also propose a definition of a smart environment. A smart environment relies on sensory data from the real world. It perceives the environment using the sensors, uses this information to reason about the environment and takes the adapted actions. This definition is similar to the concept of Smart Home seen as a closed loop system of Fig. 0.3.

Based on the different definitions of AAL and Smart Home existing in the literature and briefly recalled in this section, a definition of a Smart Home is proposed in this thesis and is given below.

**Definition 4** (Smart Home). A Smart Home is a home equipped with sensors and actuators. It is also often composed of a communication network connecting the key electrical devices and allowing them to be remotely monitored or controlled. Based on the information given by the sensors, the actuators can be controlled in order to improve comfort or to guarantee the safety of the inhabitants.

This definition is strongly related to the representation of a Smart Home given in Fig. 0.3. Note that, even if the most critical applications are relative to safety and health, thus mostly dedicated to elderly, people from other classes of age may also benefit from a comfortable and safe home.

### 1.1.2. Acceptance of home automation by the inhabitants

One of the key aspects for the installation of home automation in private houses is the acceptance of such technologies by the inhabitants. The choice of the type of sensors and of their positions as well as the usage of the data they are transmitting should be as respectful of the privacy of the inhabitants and as less invasive as possible. Several studies were conducted in order to evaluate how well this home automation is accepted by the inhabitants and most of them give recommendations to be compliant with the user needs and wills.

In the University of Kaiserslautern, Germany, a study of the acceptance of AAL technologies was conducted and the results are detailed in (Grauel and Spellerberg, 2008) and summarized in (Floeck, 2010). One conclusion of this study is that the majority of the interviewed elderly show positive attitudes towards technology. High rates of respondents can imagine using assisted living devices in the future. They however have concerns regarding the cost of such technologies. Moreover, the use of ambient technology has been judged by the interviewed people as desirable because it is less stigmatizing than devices that have to be worn on the body. The authors also highlight some challenges in the field of AAL. For instance, the monitoring of health status using home automation devices has to prove to be reliable. Low costs of installation of such devices are also of importance. Finally, the installation of sensors or actuators in the house should be performed without significant changes in the environment of the inhabitant because elderly people are often afraid of changes and rarely accept structural alteration works. Of course, the authors mainly focused on elderly people because it was of particular interest for their study and because such AAL and home automation technologies are mainly devoted to them. However, it can be imagined that home automation would be accepted by younger people in a similar way.

In another study presented in (Beringer et al., 2011) the authors found that even if a high degree of acceptance regarding AAL was observed, such technologies have the potential to profoundly affect, both positively and negatively, participants meaning and experience of the home environment. The four themes that emerged from the interview of the participants were independence, security, privacy and interference with current lifestyle. These results are similar to those from the previously reviewed study. However, the authors give some limitations to their work, particularly concerning the small number of interviews they conducted.

In addition to these studies of the acceptance of home automation technologies, ethical considerations on these technologies are given in (Borges et al., 2008). According to the authors, while processing data from a Smart Home and using them, data must be:

- Fairly and lawfully proceeded

- Processed for limited purposes

- Adequate, relevant and not excessive

- Accurate

- Not kept longer than necessary

- Processed in accordance with data subject's rights

- Secure

- Not transferred to other countries without adequate protection.

Of course, these ethical concerns are of primary importance for the users. Even if the sensor information is processed in order to detect health problem or to ensure the safety of the inhabitants, this information must remain as private as possible and should never be accessible to not concerned people.

Consequently, one of the fields of research is also to ensure the privacy of these data. For instance, in (Park et al., 2013), a comparative study of privacy protection methods for Smart Home environments is presented. The authors' conclusion is that a tradeoff must be found between privacy, latency and energy efficiency for the transmission of the data.

To conclude this section relative to the definition of Smart Home and AAL and to the acceptance of such approaches by the user, several key points can be highlighted. According to the users, a Smart Home should be reliable, low-cost, not impacting their independence, improving the security, respecting the privacy, not much inferring with current lifestyle and respecting ethical considerations. In the following section, a review of existing applications of the AAL and Smart Home field is given.

## 1.2. Different applications

As mentioned in Subsection 1.1.1, the AAL and Smart Home applications can mainly be divided into three classes: comfort, safety and health. They are detailed in the next subsections.

### 1.2.1. Comfort monitoring

The improvement of the comfort of the inhabitants may be related to different notions of comfort.

The first and widely developed application aiming to improve the comfort of inhabitants is the smart control of Heating, Ventilation and Air-Conditioning (HVAC) actuators thus leading to an adequate temperature and humidity of the environment of the inhabitants according to their needs, wills or life habits. For instance, it may be possible to learn the temporal windows where at least one inhabitant is likely to be at home and thus to optimize the control of the HVAC so that the conditions in the home, when an inhabitant enters, are satisfying.

Another possibility relative to the comfort is the smart monitoring of the lights according to the presence of the inhabitants in the different rooms or zones of the house. It may consist for instance in automatically turning on the light in the room an inhabitant just entered and then automatically turning off the light in the room the same inhabitant just left, if the room is now empty. Another application of smart monitoring of the lights, dedicated to the elderly, consists in detecting the inhabitant getting up in the night to go to the toilets and automatically turning on the lights from the bedroom to the toilets, thus indicating him the path between the two rooms and preventing him from hurting himself or to taking a wrong direction.

In a similar direction, some applications are devoted to the smart monitoring of the TV (choice of the channel or of the program) or of the music (choice of the music's style or favorite songs) in the different rooms according to users' preferences. Once again, the presence of a particular inhabitant in a room should be detected in order for the monitor to display the favorite TV show or the favorite music of this inhabitant in this room.

In this domain, the proposition and delivery of adapted services is also a field of research (Lankri et al., 2008; Lankri et al., 2009; Allègre et al., 2012). In this case, based on the location or the activity of the inhabitant, an *ad hoc* device (touchscreen, smartphone) may propose to him some services related to his current location or activity. For instance, an inhabitant entering the bedroom in the evening may be suggested to remotely close the blinds. The service *close the blind* is thus displayed and the user may or may not accept it and activate it. Such approaches can be particularly useful for elderly with memory impairments.

Furthermore, in addition to improving the comfort of the inhabitants, such approaches also contribute to save money while providing the user with a smart monitoring of energy. Indeed, the HVAC system or the lights are turned on only when required and their usage is thus optimized.

### 1.2.2. Safety ensuring

There are several ways to ensure the safety of the inhabitants. For instance, in (Floeck, 2010), two different types of threats were identified: internal and external threats.

The internal threats are for instance the occurrence of a fire or water flooding. To avoid them, most of the time specific sensors have to be installed (a smoke detector for instance). However, using simple sensors and knowing the location of each inhabitant may also contribute to ensure their safety. For instance, some approaches are aiming to automatically shut down the dangerous devices when the house is empty (i.e. all the inhabitants are outside) or to automatically turn off the faucet in the bathroom if water is flooding and nobody is present in this room (or at least to warn the inhabitants that there is a potential water flooding in the bathroom).

The external threats are for instance the presence of burglar trying to break into the house. Using the sensors inside the house, it may be possible to detect them, take a picture or record a video of them, and automatically warn the police.

### 1.2.3. Health problems detection

One of the major applications of Smart Home technologies is the detection of potential health problems. According to (Floeck, 2010), these domestic emergencies and medical problems can be ranked considering three different severity levels. An overview of this classification is given in Table 1.1. The most severe health condition concerns heart attack or stroke. Such health problems require an immediate (within minutes) intervention of the emergency services. This constitutes the high severity level (A). Less severe are the falls or unconsciousness not threatening the life of the inhabitant. For these ones, an intervention is required as soon as possible by caregivers, neighbors, friends but not necessary by emergency services. This constitutes the medium severity level (B). The low severity level (C) concerns the long term diseases. No intervention of the emergency services or caregivers is required at short term but the person should see a doctor within days to weeks (or the medical file should be updated). Such diseases are for instance dementia, hypertension or a decrease of activity.

Table 1.1.: Severity levels of domestic emergencies or medical conditions for AAL (Floeck, 2010)

| Severity level | Description | Life-threatening | Examples | Action required |
|---|---|---|---|---|
| A high | Sudden, genuine emergency | Yes | Heart attack Stroke | Immediately (within few minutes) |
| B medium | Sudden, genuine emergency | No | Fall Unconsciousness | As soon as possible (within minutes to few hours) |
| C low | Onset or exacerbation of a chronic illness | No | Hypertension Dementia Decrease of activity | Subject should see a doctor (within days to weeks) |

For the health problems of type A or B, a procedure to generate alarms should be set up. For instance, in (Floeck, 2010), a five stage procedure is proposed. This procedure is recalled in Fig. 1.2. The first step consists in the detection of the health problem. It could be done automatically through a detection approach based on the information given by the sensors of the house (as some approaches detailed in the following) or it could be a manual information given by the user pressing a button for instance. The second step is performed automatically by the computer (this computer is named PAUL in this particular Smart Home, details about PAUL are given in (Floeck, 2010)) which calls the user in the flat in order to check if the alarm is a true positive or a false alarm. If the call is answered, the user is asked to press a button on the computer to cancel the alarm. If the button is not pressed, stage 3 is reached and the alarm is forwarded to the emergency call center. The fourth step consists in someone being in the emergency call dispatcher trying to call the person again. If the person answers and depending on his answer, either the alarm is cancelled or stage 5 is reached. In this last stage, depending also on the detected health problem, friends, neighbors or caregivers are notified or the emergency services are sent to the home. This whole procedure allows being tolerant to the possible false alarms that may occur due to sensor faults or particular not yet envisaged human behaviors.

The detection of health problems of the different types (A, B or C) relies on different methods and technologies. Some of them are described in the following.

## Type A

For the health problems of type A, the continuous monitoring of certain physiological parameters is mandatory (for instance the pulse, to detect a heart attack). Thus, wearable sensors are the most appropriate to detect type A health problems and could probably not be replaced by ambient sensors without extending the delay of detection in such a way that the emergency services would not take action quickly enough. Such a wearable sensor is proposed for instance in (Anliker et al., 2004) where some physiological parameters (ECG, blood pressure, pulse)

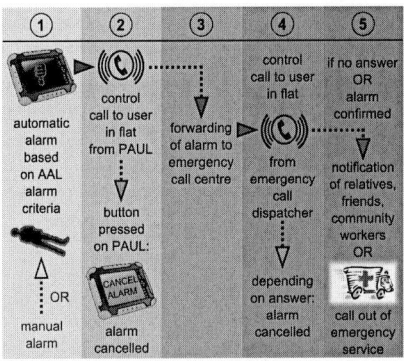

Figure 1.2.: Schematic representation of the five-stage alarm scheme (Floeck, 2010)

are monitored.

## Type B

For medium health problem of type B, different applications are possible, including the detection of the fall, the monitoring of the global inactivity or the monitoring of the duration of stay in different zones.

For instance, several methods are devoted to detect the occurrence of a fall of one inhabitant. Such approaches are mostly based on the monitoring of video images, identifying the shape of a person and trying to figure out if the person is standing or lying and to detect a sudden change from standing to lying position (Schulze et al., 2009; Rougier et al., 2011). Details on the methods, the principles and the evaluation criteria of fall detection approaches are also proposed in (Noury et al., 2007).

Other approaches are considering that medium health problems are characterized by a more or long delay of inactivity of an inhabitant. An approach aiming to monitor the global inactivity in a Smart Home is proposed in (Floeck and Litz, 2009). The details on this approach are given in Section 1.3.

In addition to considering the inactivity delay, the duration of the stay in zones of particular interest (toilets, bathroom) may also be symptomatic of a health problem (a fall or unconsciousness for instance). An approach to monitor such durations of stay is proposed in (Floeck et al., 2011) and also detailed in Section 1.3.

## Type C

In order to detect long term diseases (i.e. health problems of type C), one possibility is to monitor the long term trend in terms of Activities of Daily living (Fleury et al., 2010; Vacher et al., 2010) or of the overall level of activity (Noury et al., 2008; Noury et al., 2009; Noury et al., 2010). Activities of daily living (ADLs) refer to daily activities performed by an inhabitant within his Smart Home. Classical ADLs are (among others) preparing a meal, dining, watching TV, taking a nap, cleaning, bathing, etc... Several ADL recognition approaches are detailed in the next section. On the contrary, the overall level of activity is a unique index estimating if the inhabitant is active of not, no matter the kind of activity. Detecting a change in these long term profile may be symptomatic of a long term disease impacting the habits and lifestyle of an inhabitant.

For instance, the particular case of dementia can be detected through a change in the ADL. Monitoring the ADL may also be a method to help people suffering from dementia by providing them with reminders (Das et al., 2011; Cook, 2006; Pollack et al., 2003; Kautz et al., 2002). For instance, people starting to perform one particular ADL and forgetting what to do next (for instance, preparing a meal). The ADL as well as the current step performed by the inhabitant can be detected and the next step to perform in order to complete the ADL can be prompted to him, if he seems to have forgotten what to do next.

The aim of all these approaches (comfort, safety or health problem detection) could also be seen as to improve the trajectory of functional decline (Skubic et al., 2009). As shown in Fig. 1.3, Smart Home technologies are aimed to maintain the functional ability as long as possible. This can be done by detecting as quickly as possible the health problem or by helping the inhabitants in their everyday life (reminders for people suffering from dementia, suggestion of relevant services).

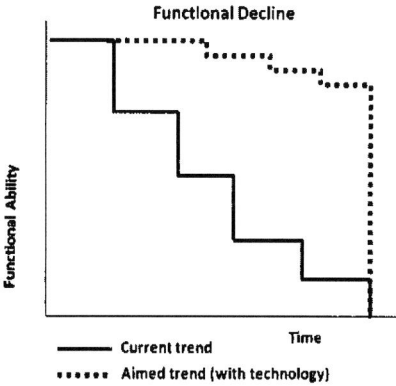

Figure 1.3.: The trajectory of functional decline (Skubic et al., 2009)

As seen in this section and to conclude, AAL applications cover a large spectrum, from comfort monitoring to health problem detection, via safety ensuring. To achieve the different goals of the different applications, different approaches have been developed, the main ones being detailed in the following section.

## 1.3. Different approaches

In order to achieve the different applications previously described, different approaches have been developed. Among these approaches, the most important is probably the recognition of Activities of Daily Living. Activity and inactivity monitoring as well as Location Tracking are also major fields of research in this area.

### 1.3.1. Activities of Daily Living recognition

As previously detailed, the ADL recognition approaches are very often used in order to improve the comfort of the inhabitant, his safety or the detection of health problem. ADL recognition consists in identifying the activities performed by one or several inhabitants using only the information provided by the sensors. Several methods exist to perform ADL recognition; some of them are briefly reviewed below.

Most of the approaches are based on machine learning techniques and are aimed to classify the different ADL starting from a set of sensor data. To do so, several techniques are proposed in the literature. For instance, in (Fleury et al., 2010), it is proposed to use Support Vector Machines (SVMs) for the classification of the ADLs. In (Nazerfard et al., 2011), association rule mining is used in order to discover temporal relations of daily activities. In (Sehili et al., 2012), a combination of two methods, GMM (Gaussian Mixture Models) and SVM (Support Vector Machines) for sound/speech discrimination and sound classification (and thus ADL recognition) is given. An event-driven approach for activity recognition using EARL (Event-Driven Activity Recognition Language) is proposed in (Storf et al., 2009). The temporal evidence theory is proposed as a solution to recognize ADL in (McKeever et al., 2010). Semi-Markov models are also a possibility, according to (van Kasteren et al., 2010). Finally, a hybrid probabilistic neural model is described in (Yan et al., 2011). The details of all these approaches are given in the relative papers; the conclusion drawn from this review is that several possible approaches exist to perform ADL recognition, mostly based on machine learning approaches.

The aim of all the previously recalled approaches is to recognize ADL based on a set of data. However, the performance of such methods has to be evaluated. In (Chikhaoui et al., 2011) and (Abidine and Fergani, 2012), the use of a confusion matrix is proposed as a solution. Particularly, in (Abidine and Fergani, 2012), a comparison of four classifiers for ADL recognition is given using a confusion matrix. It allows giving a criteria relative to the recognition accuracy of the approach for common ADLs. However, such a performance evaluation criteria requires the data to be annotated in order to enable the comparison between the real ADL (really performed by the inhabitant) and the estimated one (estimated by the considered ADL recognition approach).

### 1.3.2. Activity monitoring

For some applications, the detail of the activities performed by the inhabitants is not necessary. An overall estimation of their level of activity is sufficient. This can be done for instance by identifying human activity using video camera and detecting depth silhouettes (Jalal et al., 2011). In (Noury et al., 2008), it is proposed to use motion detector to monitor the activity of the user and in (Noury et al., 2009; Noury et al., 2010) a unique electrical index is used. Based on this electrical information, a profile of activity is obtained and can be analyzed. An example of such a profile and its analysis is given in Fig. 1.4.

### 1.3.3. Inactivity monitoring

In opposite to the previously proposed approaches aiming to monitor the activity, some authors think that monitoring the inactivity of an inhabitant is more symptomatic of particular health problems (fall, unconsciousness). Thus, in (Poujaud et al., 2008) an approach aiming to monitor the inactivity is proposed. For instance, this inactivity is shown in Fig. 1.5 (a) according to the period of the day (morning, afternoon, evening or night). This gives the

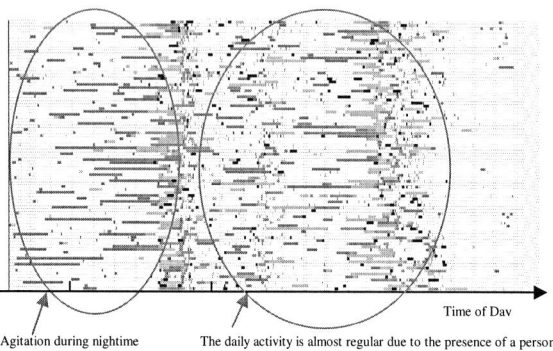

Figure 1.4.: Spatiotemporal diagrams for a person suffering from Alzheimer. The person has a regular activity during the day due to an accompanying person and is very agitated during nighttime. (Noury et al., 2009)

lifestyle profile and describes the habits of the inhabitant, and thus can be monitored in order to detect brutal change.

Another approach for inactivity monitoring is proposed in (Floeck and Litz, 2009). The inactivity is monitored along the day and profiles are obtained. An example of such a profile is given in Fig. 1.5 (b). Based on this profile, learned during a certain duration (one week, one month or more), a threshold (static or more complex) can be defined and the inactivity is then monitored. If exceeding the threshold, an abnormal inactivity is detected, symptomatic of a health problem, and an alarm may be triggered. As it can be seen in the figure, in order for the approach to work well, it has to be known if the inhabitant is really inactive or if the inactivity comes from the inhabitant being outside the house. This will be detailed in the next subsection.

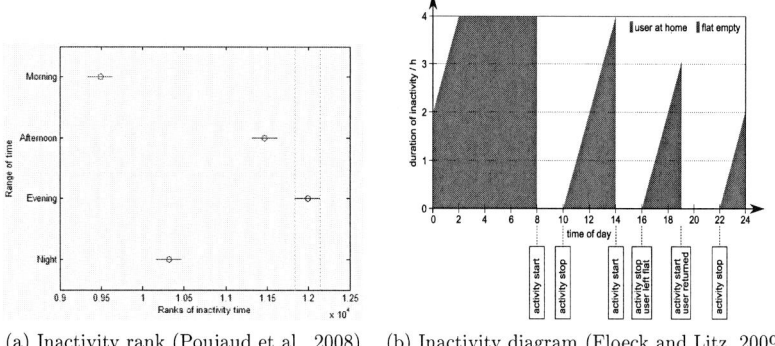

(a) Inactivity rank (Poujaud et al., 2008)   (b) Inactivity diagram (Floeck and Litz, 2009)

Figure 1.5.: Profiles of inactivity (Poujaud et al., 2008; Floeck and Litz, 2009)

Moreover, a comparison between activity and inactivity monitoring is given in (Floeck et al., 2012). Monitoring activity and inactivity have different objectives and allow different health problems to be detected (for instance dementia using activity monitoring, fall using inactivity monitoring). Of course, activity and inactivity are strongly related.

### 1.3.4. Location Tracking

Finally, the last approaches to improve comfort, safety or health are using the real-time Location Tracking of the inhabitants. They are based on different techniques and have different aims. Some of them are summarized below. It is also shown that using Location Tracking improves some of the other approaches (ADL recognition, inactivity monitoring).

#### Different aims and techniques for indoor Location Tracking

A simple aim for Location Tracking is to estimate the home occupancy i.e. if at least one inhabitant is inside the house or if the house is empty. An approach based on an intelligent agent is proposed in (Makonin and Popowich, 2011) to determine the home occupancy. Another approach is proposed in (Floeck, 2010), based on a Finite State Machine called "Presence Automaton" given in Fig. 1.6. This automaton is used online and after each observed sensor event or after a certain delay, the active state is updated, thus giving an estimation of the actual home occupancy (flat occupied, flat unoccupied or presence unclear). Note that the occupancy estimation is inaccurate when the active state is "presence unclear".

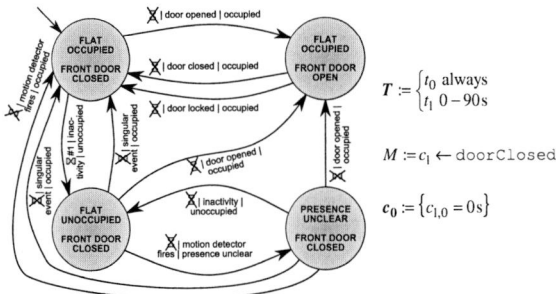

Figure 1.6.: Finite state machine (automaton) for determining human presence inside the flat (Floeck, 2010)

In (Bahl and Padmanabhan, 2000) and (Priyantha et al., 2000), Location Tracking approaches based on Radio Frequency sensors are proposed. In (Zhang et al., 2011), an approach based on RFID sensors is also proposed. Such approaches are mostly based on triangulation techniques in order to obtain a more or less accurate estimation of the position of the user. This can be used in Smart Homes with quite good accuracy. The given result is the position of an inhabitant which could obviously be post-treated in order to obtain the room in which the inhabitant is.

Several approaches are also reviewed in (Hightower and Borriello, 2001). The authors proposed a survey of the Location Tracking techniques and sensors. They are mostly based on wearable sensors and triangulation or proximity detection approaches.

In (Rahal et al., 2008), the usage of IR presence detectors, tactile carpets, smart light switches, electric contacts on doors and pressure detectors is proposed in order to perform Location Tracking of a single inhabitant. The Location Tracking algorithm is not detailed but the authors propose to take into account the potential sensor faults (wrong information transmitted by the sensors). These faults are avoided using sensor fusion (in this case Bayesian filtering). The paper shows the efficiency of such approaches to perform fault-tolerant Location Tracking. In order to evaluate the accuracy of the location estimation, a video camera and the tactile carpets information are considered as giving the real location (in which zone the inhabitant is) to be compared with the estimated location given by the other sensors and the Bayesian-filtering algorithm. According to the authors, the results are convincing. Moreover, since it is profile-independent, this approach can easily be deployed in the future homes that are being conceived in the laboratory. Furthermore, in the mind of the authors only ambient and unobtrusive sensors should be implemented in order to perform Location Tracking.

In (Roy et al., 2003), a Location Tracking approach is proposed and used for location aware resource management in Smart Homes (i.e. related to the comfort). Although the paper focuses more on the smart resource management aspect, a Location Tracking approach is proposed, based on two graphs, one representing the connectivity of the zones of the house (see Fig. 1.7) and one representing the connectivity of the sensors of the house (see Fig. 1.8).

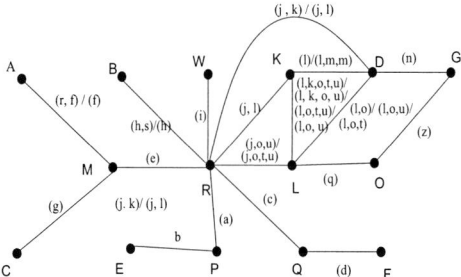

Figure 1.7.: Graph Representing the Connectivity of Zones/Locations (Roy et al., 2003)

In the graph representing the connectivity of zones (see Fig. 1.7), each node represents one zone of the house and each edge between two zones is labeled with the name of a sensor between the related two zones. Edges are not oriented and note that there are no self-loops (i.e. edges having the same node for its source and its destination).

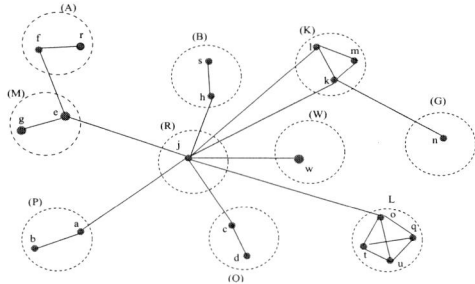

Figure 1.8.: Graph Representing the Connectivity of In-building Sensors (Roy et al., 2003)

In the graph representing the connectivity of the sensors (see Fig. 1.8), each node represents a sensor and each edge represents the connectivity between individual sensors. Moreover, the position of each sensor can be represented; the name of the related zone is given near to the groups of sensors, circled with a dashed line.

Despite these interesting representations, the algorithm for Location Tracking is not given.

In addition to the previously described works, the same authors proposed in (Das et al., 2006) to extend their Location Tracking approach to the case of multiple inhabitants. However, the paper is more focused on the context-aware resource management approach.

Approaches also exist for Location Tracking based on video cameras. For instance, in (Yu et al., 2006), an approach for human localization via multi-cameras and floor sensors is proposed.

Finally, to conclude this review of Location Tracking approaches, in (Floeck, 2010), the author proposes an approach for Location Tracking based on a Finite State Machine given in Fig. 1.9. Based on such a model, coupled with the presence automaton given in Fig. 1.6 and the home occupancy estimator, the room in which the inhabitant is located is estimated at each time. However, the author gives no indication on how to build such an explicit model for a given apartment and he does not deal with the problem of multiple inhabitants in the house.

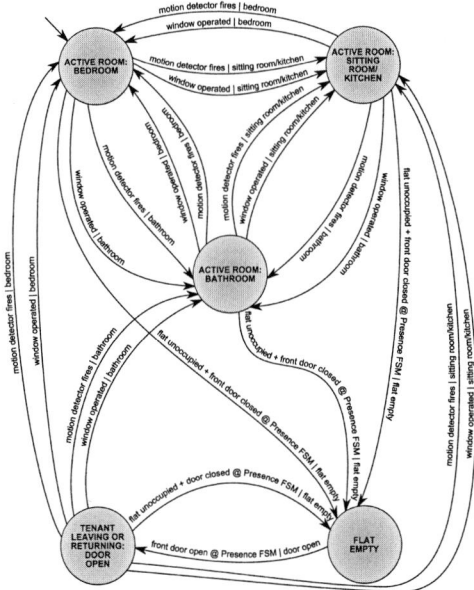

Figure 1.9.: Finite State Machine tracking the current location occupied by the user (Floeck, 2010)

To conclude, there are many different Location Tracking approaches, most of them are based on body-worn sensor and triangulation techniques. However, approaches based only on ambient sensors also exist and some of them are based on an explicit model.

**Location Tracking improves other approaches**

As seen before, ADL recognition, activity and inactivity monitoring are based on the sensors information. However, some of these approaches show better performances when the information of the current location of the inhabitants is also taken into account.

This is particularly the case for ADL recognition. It has been shown in (Chen et al., 2012; Lu and Fu, 2009; Seth Long and Holder, 2011; Wilson and Atkeson, 2005; Liao et al., 2005) that the usage of Location Tracking improves the results of the developed ADL recognition approaches.

Moreover, Location Tracking improves inactivity monitoring by considering home occupancy. The home occupancy estimation method of (Floeck, 2010) is used in order to determine if the observed inactivity is due to a real inactivity of the inhabitants (symptomatic of a health problem) or to the home being empty (see an example of the two cases in Fig. 1.5 (b)). In the case of the home being empty, no alarm should be raised if an exceeding inactivity is detected because it will surely be a false alarm. In addition, note that such a home occupancy estimation procedure should be particularly reliable because if an inhabitant being really inside the house and having a health problem leading to a long inactivity is estimated to be outside, no alarm will be raised.

Finally, indoor Location Tracking also improves inactivity monitoring by considering local inactivity profiles. For instance, in (Poujaud et al., 2008) the inactivity in each room is considered (see Fig. 1.10). Based on this differentiation of the observed inactivity within the different rooms, the result of the monitoring shows better performances. Local thresholds for detecting an abnormally long inactivity are smaller in specific rooms (bathroom and WC, according to Fig. 1.10) than a unique threshold for the whole home. It is thus possible to drastically reduce the delay of detection in these particular rooms.

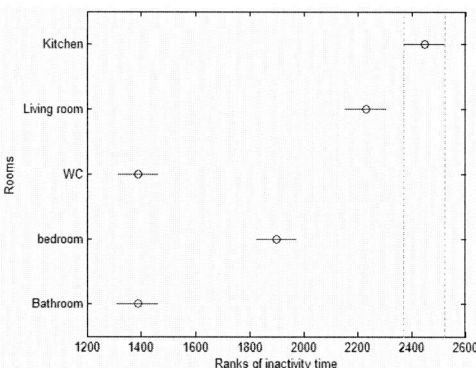

Figure 1.10.: Variance and mean ranks of inactivity according to the room (Poujaud et al., 2008)

The same holds when considering the duration of stay in each room as proposed in (Floeck et al., 2011). The duration of stay in each room is monitored (see Fig. 1.11), and based on this observed profile, a different threshold for abnormally long duration of stay can be set for each room. As for the previously described approach concerning the inactivity, the delay of detection of an abnormally long duration of stay in some rooms is drastically reduced, particularly in the bathroom.

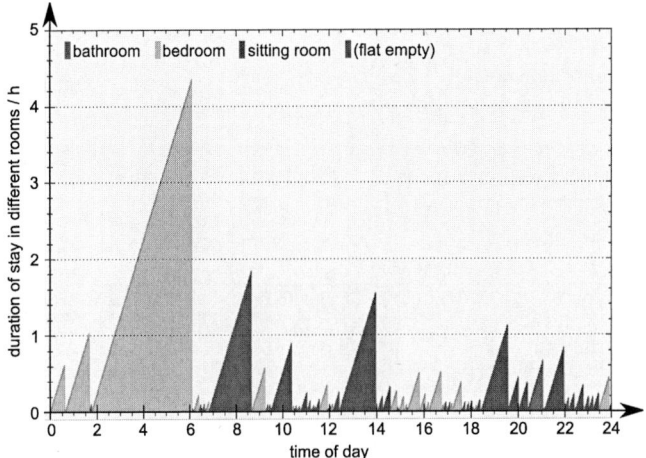

Figure 1.11.: Duration of stay in various rooms throughout the day (Floeck, 2010)

This overview of the existing approaches for Location Tracking is not exhaustive but is assumed to show a pertinent review of the state of the art in this domain. In the following subsection, some limits of these existing approaches are given.

### Limits of the currently developed Location Tracking approaches

The first limit concerns the technological choices. Most of the Location Tracking approaches are based on the use of wearable sensors (mostly RFID). As written in the section relative to the acceptance of home automation by the user, wearable sensors are seen as being intrusive and a user may forget to wear it. This last problem (forgetting to wear a sensor) may lead to a false result of Location Tracking, which can have dramatic consequences for applications (particularly relative to health) requiring a strongly reliable Location Tracking result.

Most of these approaches also suffer of the lack of an explicit-model. Even if a model exists, it is often strongly depending on the inhabitant habits. There is one model for each inhabitant or one for a given set of inhabitants. Thus, no general model can be given for a given Smart Home, whoever the inhabitants are. If the same Smart Home is constructed several times with the same topology and the same instrumentation, a particular model has to be learned for each one of them.

In addition to this limit relative to the model, no model-building methodology is proposed. Thus, someone willing to apply the existing approaches to his own Smart Home may have difficulties to build the model and to evaluate its correctness.

## 1.4. Different sensors

To conclude this review of the applications and techniques of AAL and Smart Home, a review of the different sensors usually used in Smart Homes is proposed. A table of existing sensors and approaches is already proposed in (Cardinaux et al., 2011), however, being not as complete as the two tables proposed below. Each table is summarizing the existing sensors of each class, respectively wearable and ambient sensors.

### 1.4.1. Wearable sensors

An overview of the existing wearable sensors and their applications is given in Table 1.2. There are two main sorts of wearable sensors, the first ones measuring a physiological parameter (body temperature, blood pressure, pulse), and the other ones relative to the position or the movement of the user (accelerometer, GPS, RFID, Ultrasonic badges).

Table 1.2.: Review of existing wearable sensors

| Type of sensor | Applications for Smart Home and Ambient Assisted Living | | | | | | |
|---|---|---|---|---|---|---|---|
| | Comfort | Safety | Health Problem | | | ADL recognition | Location Tracking |
| | | | (A) | (B) | (C) | | |
| Body temperature | | | | x | x | | |
| Blood pressure, Pulse | | | x | x | x | | |
| Accelerometer | | | | x | x | x | |
| GPS | | | | | | | x |
| RFID | | | | | | x | x |
| Ultrasonic badges | | | | | | | x |

### 1.4.2. Ambient sensors

An overview of the existing ambient sensors and their applications is given in Table 1.3. These sensors are sorted from the most intrusive (camera, microphone) to the less intrusive (simple indicator of the overall energy consumption in the home).

Table 1.3.: Review of existing ambient sensors

| Type of sensor | Applications for Smart Home and Ambient Assisted Living | | | | | | |
|---|---|---|---|---|---|---|---|
| | Comfort | Safety | Health Problem | | | ADL recognition | Location Tracking |
| | | | (A) | (B) | (C) | | |
| Camera | x | x | | x | x | x | x |
| Microphone | x | x | | x | x | x | x |
| Smart Floor | x | | | x | x | x | x |
| Motion Detector | x | x | | x | x | x | x |
| Door Barrier Sensor | | x | | | | | x |
| Bed Sensor | | | | x | x | x | x |
| Contact sensor (Door, Windows) | | | | | | x | |
| Ambient temperature | | x | | | | | |
| Water consumption | x | x | | | x | x | |
| Energy consumption | x | | | | x | x | |

## 1.5. Problem statement of explicit-model-based Location Tracking

Based on the previously given state of the art, it has been decided to focus on indoor Location Tracking of single or multiple inhabitants in Smart Homes. This thesis proposes a whole new approach to deal with this objective. The review of the existing approaches leads to several considerations and assumptions, detailed in this section. Finally, based on these considerations and assumptions, a new formulation of the problem of Location Tracking is proposed.

One of the main concerns when dealing with smart environments is the respect of the user's habits and privacy. This leads to two considerations relative to the sensors that should be used in Smart Homes:

- The user should accept the observation of his every movement. To help him towards this acceptance, only non-wearable sensors (i.e. **ambient sensors**) should be used. Using no wearable sensors also allows not having to consider the problem of an inhabitant forgetting to wear it or the sensor falling down and not being worn anymore.

- Even if using only ambient sensors, the respect of the user's privacy has still to be ensured. To do so, in this thesis, it is chosen to use **the less intrusive sensors**. Thus, no camera but only motion detectors or floor pressure sensors for instance are used.

Considering also a financial point of view, any proposed solution should be as less expensive as possible. This leads to the following consideration on the sensors:

- The use of **low-cost sensors** is strongly encouraged. This consideration is not contradictory to the two first ones, because considering only ambient and non-intrusive sensors often leads to consider only low-cost sensors.

Based on these three considerations, only **binary sensors** or eventually sensors delivering a signal that can be interpreted as being binary using thresholds are considered in this thesis. In addition, an assumption is made on the potential occurrence of sensor faults:

- It is assumed that all the sensors are always operating in a **fault-free** manner (or at least in an acceptable manner) i.e. the functioning of the sensors does not impact the detection of the motion or the presence of an inhabitant and thus should not impact the result of the Location Tracking.

Regarding the inhabitants of the house (inhabitants living there, guests or pets) and their relation to the sensors, three assumptions are made:

- The information given by the sensors does **not depend on the ability or the willingness** of the inhabitant to provide this information. For instance, if a door is equipped with a door barrier sensor and a door contact sensor, an inhabitant crossing the door will systematically be detected by the barrier sensor but will be detected by the contact sensor only if this inhabitant opens or closes the door in addition to crossing it. Consequently, in this thesis, door contact sensors will not be used. For similar reasons, switch sensors are also not considered because while entering a room an inhabitant may or may not switch the light on, depending on the sun light or his life habits.

- It is also assumed that each inhabitant has a **totally free behavior** and each inhabitant behaves independently from the other. Consequently, each inhabitant living in an instrumented environment is seen as a spontaneous event generator. These events are the rising and falling edges of the signal emitted by each binary sensor of the house.

- Considering the topology of an apartment and a potential lack of instrumentation in some areas, the assumption of **partial observation** of the behavior of each inhabitant is also made.

Finally, since the aim of this thesis is to propose an approach for model-based Location Tracking, a last assumption is made:

- **No model of the human behavior** is available. It is assumed that this model is hard to build and is overly depending on particular behavior of particular inhabitant. The aim of the proposed approach is to be usable whoever the inhabitants are.

Based on these considerations, the problem of online model-based Location Tracking can be reformulated in terms of a Discrete Event System (DES) problem: how to **estimate** in real time the **current location** of the inhabitants, considered as **spontaneous event generators**, based on a potentially incomplete observed **sequence of sensor events**?

Thus, based on the description of a Smart Home proposed in Fig. 0.3 of the introductory chapter, the Location Tracking is introduced as shown in Fig. 1.12.

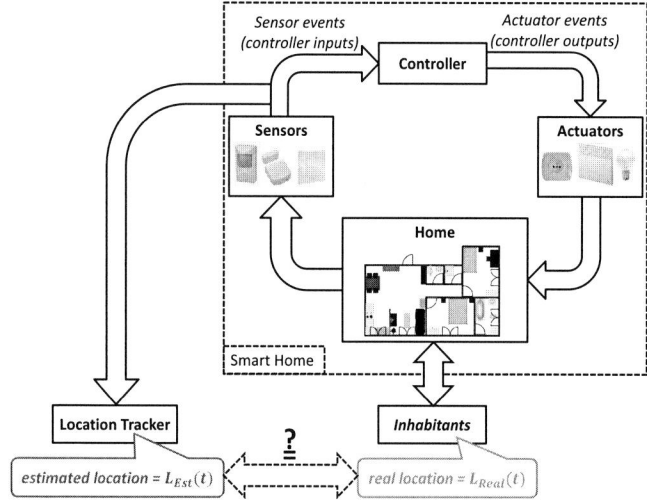

Figure 1.12.: Concept of online Location Tracking in Smart Homes

Sensor events are read online and are all considered as being observable. The sensors are all considered as being operating in a fault-free manner. The estimation of the current location of the inhabitants (named $L_{Est}$ in the figure) is updated each time a new sensor event is observed. Of course, the estimated location should be the most accurate i.e. as close to the real location of the inhabitants (named $L_{Real}$ in the figure) as possible.

## Conclusion

In this chapter, a detailed state of the art in the fields of Ambient Assisted Living and Smart Homes has been given. Based on the review of the existing applications, techniques and sensors as well as considerations on the user needs, the problem of model-based Location Tracking in Smart Homes has been stated and particularly, it has been expressed in terms of a DES problem. Moreover, assumptions and considerations that will hold in the following of this thesis have been detailed.

As expressed in the introduction of this thesis (see Fig. 0.4), the proposed approach is divided in three main steps which are **modeling**, **Location Tracking** and **performance evaluation**. In the next chapter, the procedure to model the detectable motion of one or several inhabitants living in a Smart Home is given. This constitutes the first step toward model-based Location Tracking.

# 2. Modeling of the detectable motion of inhabitants in a Smart Home

## Introduction

Based on the previous review of the state of the art, there is a lack of formal model for the different methods aiming at Location Tracking. In this chapter, it is proposed to use Finite Automaton, a formalism of the Discrete Event System theory, to model the detectable motion of a single inhabitant in a first time and of multiple inhabitants in a second time. The aim of these models is to perform, in a future step, model-based Location Tracking. A unique formalism, the Finite Automaton, is proposed as a tool to model the detectable motion of inhabitants in a Smart Home but different procedure leading to this model are detailed in this chapter and compared on similar case studies.

## 2.1. Detectable motion of a single inhabitant

Since the Location Tracking of inhabitants can be expressed as a problem of Discrete Event System modeling, an appropriate way to model the detectable motion of an inhabitant is to use Finite Automata (Danancher et al., 2012). The following definition of a Finite Automaton (FA) has been proposed in (Cassandras and Lafortune, 2009).

**Definition 5** (Finite Automaton (Cassandras and Lafortune, 2009)). A *Finite Automaton* is a quadruplet $Aut = (Q, \Sigma, \delta, Q_0)$ with:

- $Q$ a set of states,

- $\Sigma$ an alphabet of events,

- $\delta : Q \times \Sigma \to 2^Q$ the transition function,

- $Q_0 \subseteq Q$ the set of initial states.

Note that using this definition, the automaton may be non-deterministic. It may have several initial states and from one state $q$ on the occurrence of an event $\sigma$, there may be more than one destination state (according to the definition of the transition function $\delta$).

In addition, the following notation is defined: $\delta(q, \sigma)!$ means that $\delta(q, \sigma) \subseteq Q$ i.e. at least one transition from state $q \in Q$ labeled with the event $\sigma \in \Sigma$ is defined ($\delta(q, \sigma) \neq \emptyset$).

This formalism is used in this work to represent the detectable motion of an inhabitant in a home instrumented with sensors. The following definition is proposed.

**Definition 6** (Detectable Motion Automaton). A *Detectable Motion Automaton* ($DMA$) is a Finite Automaton representing the different possible locations of a single inhabitant and the possible observation of his change of location.

Formally, $DMA = (Q, \Sigma, \delta, Q_0)$ with:

- $Q$ the set of states. Each state represents a location of the inhabitant i.e. each state represents a zone of the considered home.

- $\Sigma$ an alphabet of events. Each event represents either a rising or a falling edge of a binary sensor. By convention, in this thesis, the rising edge of a sensor $s$ is denoted $s\_1$ and its falling edge is denoted $s\_0$.

- $\delta : Q \times \Sigma \to 2^Q$ the transition function. Each transition represents an observable change of location of the inhabitant. Note that self loops (i.e. transitions having the same source state and destination state) are also considered as being representative of an observable change of location.

- $Q_0 \subseteq Q$ the set of initial states. Depending on the knowledge of the initial location of the inhabitant, there can be one or more initial states.

This $DMA$ model is not aimed to be a model of the behavior of the inhabitant because it is assumed that the inhabitant has an arbitrary and potentially irrational behavior that cannot be easily modeled. Thus it is considered that the behavior of the inhabitant is totally free and unknown. The $DMA$ is seen as a model of the observable motion of the inhabitant in the home and should be as permissive as possible in order to cope with all the possible real behaviors of the inhabitant. By doing so, the model is also not depending on the considered inhabitant and can be used for Location Tracking whoever the inhabitant is.

### 2.1.1. Overview of different approaches for modeling

In this work, different approaches are proposed to build a $DMA$. Their common point is that the resulting FA should satisfy to the definition of a $DMA$ (Definition 6). These approaches are summarized in the Fig. 2.1.

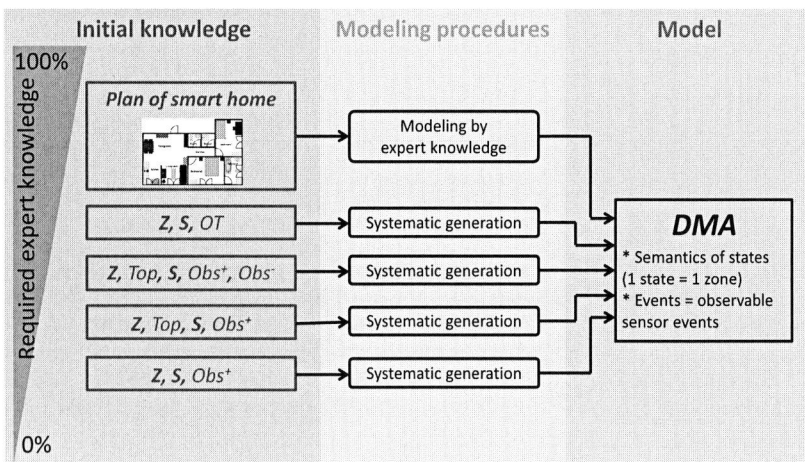

Figure 2.1.: Overview of the modeling approaches

These approaches are requiring different degrees of expert knowledge to define their inputs. On the top of the figure, the expert is required to build manually the whole model, starting from his knowledge of the type and position of each sensors, the topology of the house and possibly the habits of the inhabitant. In addition, methods involving less expert knowledge and aiming at the systematic creation of the $DMA$ are also proposed and described in the following. They are based on a more or less formalized description of the topology and the instrumentation of the apartment. These more or less formal descriptions, called $\mathbf{Z}$, $Top$, $\mathbf{S}$, $Obs^+$ and $Obs^-$ are defined in details in Subsection 2.1.4 while describing the different approaches of systematic generation.

Note that the common point of all these approaches is their output being always a $DMA$ satisfying to Definition 6.

A case study is proposed in the next section in order to illustrate all the approaches aiming to build the $DMA$, highlight their differences and compare their results.

### 2.1.2. Presentation of the case study

An apartment having an average size and planned for 2 or 3 persons living inside (for instance a couple and their child) is considered. A three dimensional view of this apartment has been created using the software Sweet Home 3D[1] and is shown in Fig. 2.2. Its topology of this apartment is organized as follow (see Fig. 2.3 (a)): there is only one front door between the outside and the dining room. This dining room constitutes an open space with the kitchen and the living room. Then, a corridor allows going from this open space to the toilets, the shower and the two bedrooms. Finally, a "private" bathroom can be reached only by crossing the second bedroom.

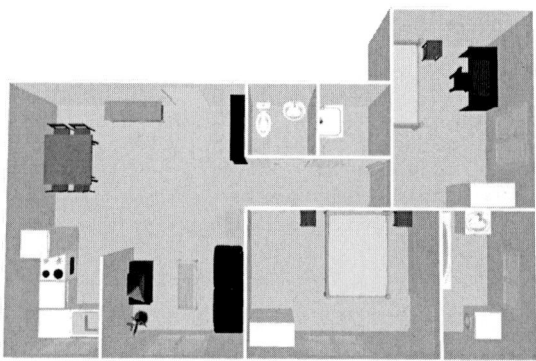

Figure 2.2.: 3D view of the Smart Home case study (using the software Sweet Home 3D)

In addition, an illustrative instrumentation is described in Fig. 2.3 (b). Five motion detectors are installed in the house, $MD_1$ is observing the whole first bedroom, $MD_2$ is observing the corridor only, $MD_3$ is observing the whole second bedroom, $MD_4$ the bathroom and $MD_5$ the whole open space i.e. the dining room, the kitchen and the living room. Two door barrier sensors are also installed, the first one $DB_1$ on the door between the toilets and the corridor and the second one $DB_2$ between the first bedroom and the corridor. Of course, there may be many more sensors (for instance floor pressure sensors in the bathroom or near the front

---

[1]http://www.sweethome3d.com

door of the house, or more motion detectors in other rooms). However, only this limited instrumentation is considered in a first time for the sake of clarity and to highlight particular points.

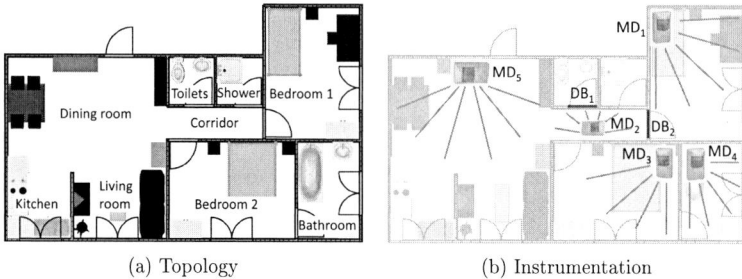

(a) Topology          (b) Instrumentation

Figure 2.3.: Topology and instrumentation of the Smart Home case study

### 2.1.3. Manual construction of a model by expert knowledge

The first idea that comes in mind to model the motion of an inhabitant is to ask a Smart Home expert to build this model. To do that, the expert may start by defining the states of the $DMA$ by considering them as being the important areas of the house in which the inhabitant is required to be accurately located when he is really there.

Then, if the expert has an idea of the initial location of the inhabitant (for instance, if the location tracker can only be turned on while the inhabitant is in the living room near the computer), then the expert can define the initial state. If he has no idea of the initial location of the inhabitant (for instance, if the location tracker can be turned on at any time remotely), then the expert can define all states as being initial.

There is *a priori* not much choice to define the events of the alphabet. Since only binary sensors are considered, their rising and falling edges constitute the events of the $DMA$. However, depending on the choice of the expert, some events may be considered as being irrelevant for Location Tracking (due to technical considerations for instance) and thus may not be included in this alphabet.

Finally, the transition function has to be defined based on the expert knowledge of the Smart Home and of the sensor technologies.

Based on this, it can be seen that there are some rules of good practice. Moreover, a Smart Home expert may not be accustomed to use the formalism of Finite Automaton to build a model. Consequently, several approaches aiming at the systematic generation of the model based on formalized expert knowledge are proposed in the following section.

### 2.1.4. Systematic generation of a model

#### Overview of the different developed approaches

Derived from the rules of good practice highlighted in the previous subsection, several approaches for systematic generation of a model were developed and are described in this subsection. Since the designer of the Smart Home may not be familiar with the FA formalism, intermediate formalisms are proposed to describe a partition of the house in different zones and its instrumentation with sensors. An algorithm is then applied in order to automatically create the $DMA$ relative to this combination zone partition - instrumentation). In the following, four different approaches are developed. They are sorted by growing required expert knowledge. Thus the first one requires as few expert knowledge as possible for systematic model building i.e. the description of the zones of the house ($\mathbf{Z}$), the description of the sensors ($\mathbf{S}$) and the zones being observed by each sensor ($Obs^+$). In the second approach, the topology ($Top$) is added to the previous descriptions, requiring a more important expert knowledge. In the third approach, a more accurate and alternative description of the zones being observed by each sensor ($Obs^+$ and $Obs^-$ instead of only $Obs^+$) is proposed. Finally, a formalism including both the topology and the observation ($OT$ instead of $Top$, $Obs^+$ and $Obs^-$) is proposed, however as it will be shown, this last approach is as hard to perform as the manual model building and thus is less interesting than the other ones.

**Approach 1, based on the least expert knowledge**

Since the aim of Location Tracking is to estimate in real time the location of the inhabitant, the zones of interest should be described by keeping this objective of Location Tracking in mind. Thus, it is proposed to split a Smart Home in different zones by respecting only one rule: the whole environment (inside and outside the Smart Home) should be covered by non-overlapping zones. The following definition of a vector called $\mathbf{Z}$ is thus given.

**Definition 7** (Zone Partition $\mathbf{Z}$). A *Zone Partition* is the partition of an environment into zones such that the whole environment is covered by zones and there are no overlapping zones. A *Zone Partition* is formally described by a vector of zones denoted $\mathbf{Z}$ where each element is the name of a zone.

Obviously, this zone partition is not unique, because, for a same Smart Home, there are several different possible zone partitions it can be chosen from. The influence of this choice is discussed in Subsection 2.1.5. However, as an example, a zone partition of the case study is considered here in order to illustrate the concept. This zone partition $\mathbf{Z}$ is shown in Fig. 2.4, the choice has been made to consider one zone for each room of the home (zones 1 to 7). In addition, one zone (zone 8) is representing the whole outside of the house.

This zone partition is formally written with the following vector $\mathbf{Z}$:

$$\mathbf{Z} = \begin{bmatrix} \mathbf{Z}_1 \\ \mathbf{Z}_2 \\ \mathbf{Z}_3 \\ \mathbf{Z}_4 \\ \mathbf{Z}_5 \\ \mathbf{Z}_6 \\ \mathbf{Z}_7 \\ \mathbf{Z}_8 \end{bmatrix}$$

Besides this zone partition, the sensors installed in the house are described. It consists directly in the list of the different sensors. The following definition of an instrumentation called $\mathbf{S}$ is given.

**Definition 8** (Instrumentation $\mathbf{S}$). An *Instrumentation* represents the sensors that are installed in the smart environment. An *Instrumentation* is formally described by a vector of sensors denoted $\mathbf{S}$ where each element is the name of a sensor.

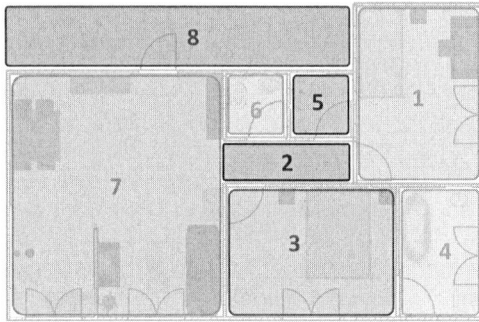

Figure 2.4.: Zone partition where 1 zone = 1 room

For the considered case study (see Fig. 2.3 (b)) there are seven sensors that could be represented by the following vector $\mathbf{S}$:

$$\mathbf{S} = \begin{bmatrix} MD_1 \\ MD_2 \\ MD_3 \\ MD_4 \\ MD_5 \\ DB_1 \\ DB_2 \end{bmatrix}$$

In addition to this description of the sensors, the link between the zone partition and the instrumentation is represented by the zones observed by the sensors, a matrix called $Obs^+$. The following definition of this matrix is given.

**Definition 9** (Zones After Observation $Obs^+$). The *Zones After Observation* related to a Zone Partition $\mathbf{Z}$ and an instrumentation $\mathbf{S}$ represent the zone(s) in which the inhabitant is after being observed by each sensor of the instrumentation. *Zones After Observation* are formally described by a ($|\mathbf{S}| \times |\mathbf{Z}|$)-matrix denoted $Obs^+$ where $Obs^+_{(i,j)} = 1$ if the inhabitant is assumed to be in zone $\mathbf{Z}_j$ after a motion is observed by the sensor $\mathbf{S}_i$, 0 else.

The matrix $Obs^+$ corresponding to the previously described zone partition $\mathbf{Z}$ and instrumentation $\mathbf{S}$ for the case study of Fig. 2.3 (b) is given below. For instance, it can be seen that the sensor $\mathbf{S}_1 = MD_1$ is observing the zone $\mathbf{Z}_1$ (i.e. the first bedroom) and this zone only because $Obs^+_{(1,1)} = 1$ and $\forall j \neq 1\ Obs^+_{(1,j)} = 0$.

The sensor $\mathbf{S}_6 = DB_1$ is observing two zones ($\mathbf{Z}_2$ the corridor and $\mathbf{Z}_6$ the toilets) because $Obs^+_{(6,2)} = 1$, $Obs^+_{(6,6)} = 1$ and $\forall j \notin \{2,6\}\ Obs^+_{(6,j)} = 0$.

Note also that some zones are observed by no sensor, for instance the zone $\mathbf{Z}_5$ (the shower) because $\forall i\ Obs^+_{(i,5)} = 0$.

$$Obs^+ = \begin{bmatrix} 1 & 0 & 0 & 0 & 0 & 0 & 0 & 0 \\ 0 & 1 & 0 & 0 & 0 & 0 & 0 & 0 \\ 0 & 0 & 1 & 0 & 0 & 0 & 0 & 0 \\ 0 & 0 & 0 & 1 & 0 & 0 & 0 & 0 \\ 0 & 0 & 0 & 0 & 0 & 0 & 1 & 0 \\ 0 & 1 & 0 & 0 & 0 & 1 & 0 & 0 \\ 1 & 1 & 0 & 0 & 0 & 0 & 0 & 0 \end{bmatrix}$$

Based on $\mathbf{Z}$, $\mathbf{S}$ and $Obs^+$, Algorithm 2.1 is proposed to systematically generate the $DMA$.

---

**Algorithm 2.1** Generation of the $DMA$ starting from $\mathbf{Z}, \mathbf{S}, Obs^+$

---

**Require:** $\mathbf{Z}, \mathbf{S}, Obs^+$
1: $DMA := \langle Q, \Sigma, \delta, Q_0 \rangle$
2: **for** $i := 1$ to $|\mathbf{Z}|$ **do**
3:     Create state $q_i = \mathbf{Z}_i$ in $Q$
4:     Set state $q_i$ initial, $q_i \in Q_0$
5: **end for**
6: **for** $k := 1$ to $|\mathbf{S}|$ **do**
7:     Create event $\mathbf{S}_k\_1$ in $\Sigma$
8:     **if** $\mathbf{S}_k$ is not a motion detector **then**
9:         Create event $\mathbf{S}_k\_0$ in $\Sigma$
10:     **end if**
11: **end for**
12: **for** $(i, j, k) := (1, 1, 1)$ to $(|\mathbf{Z}|, |\mathbf{Z}|, |\mathbf{S}|)$ **do**
13:     **if** $(Obs^+_{(k,j)} = 1)$ **then**
14:         Define $\delta(q_i, \mathbf{S}_k\_1) = q_j$
15:         **if** $\mathbf{S}_k$ is not a motion detector **then**
16:             Define $\delta(q_i, \mathbf{S}_k\_0) = q_j$
17:         **end if**
18:     **end if**
19: **end for**
20: **return** $DMA$

---

The first step consists in defining the set $Q$ of the states of $DMA$ (line 3). Each state represents a zone of the house. Furthermore, each state of the automaton is defined as being an initial state of the automaton (line 4) because the initial location of the inhabitant is assumed to be unknown. Knowing accurately the initial location is not necessary to perform online Location Tracking because the current estimation of the location of each inhabitant does not strongly depend on their initial location. If for some Smart Home applications it is mandatory to know the initial location of each inhabitant, some techniques (for instance in (Shu and Lin, 2013)) can be used to determine the initial state of an automaton after observing a more or less long sequence of events. Then, the events are created (line 6 to 10). Both the falling edges and rising edges are considered for the sensors like door barrier sensors or floor pressure sensors since both edges are indicating a movement of the inhabitant. On the contrary, only the rising edge is considered for motion detectors because a falling edge is not representative of a change of location but it can just be symptomatic of someone staying motionless in a zone. Finally, the transitions are defined (line 12 to 19). A transition exists between a state $q_i$ and a state $q_j$ labeled with the event $\mathbf{S}_k\_1$ (and a transition labeled with the event $\mathbf{S}_k\_0$ also exists for sensors not being motion detectors) if the zone related to the destination state $q_j$ is observed by the sensor $\mathbf{S}_k$ ($Obs^+_{(k,j)} = 1$). By doing so, from each state of the $DMA$ and for each sensor event a transition labeled with this sensor event is defined and the destination state is relative to the observed zones of the sensor.

By applying this algorithm on the case study described by the previous $\mathbf{Z}$, $\mathbf{S}$ and $Obs^+$, the $DMA$ of Fig. 2.5 is systematically built.

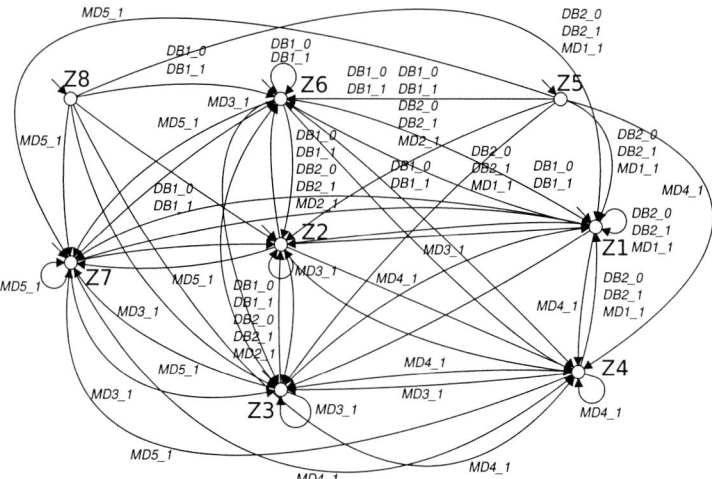

Figure 2.5.: $DMA$ generated from $\mathbf{Z}, \mathbf{S}, Obs^+$ with Algorithm 2.1 (for clarity reasons, the transitions' labels are not all shown)

The aim of this $DMA$ is to represent all the possible behavior of a single inhabitant. However, in this $DMA$ based on $\mathbf{Z}$, $\mathbf{S}$ and $Obs^+$, it can be seen that a direct motion from zone $\mathbf{Z}_1$ to zone $\mathbf{Z}_4$ is observable with the sensor event $MD_4\_1$. But, if all the sensors of the house have a fault-free behavior and if there is only one inhabitant in the house, this inhabitant moving from $\mathbf{Z}_1$ (the first bedroom) to $\mathbf{Z}_4$ (the bathroom) should be observed first in the corridor ($\mathbf{Z}_2$) by the sensor $MD_2$, then in the second bedroom ($\mathbf{Z}_3$) by the sensor $MD_3$ and finally in the bathroom ($\mathbf{Z}_4$) by the sensor $MD_4$. Consequently, the transition from $\mathbf{Z}_1$ to $\mathbf{Z}_4$ labeled with the event $MD_4\_1$ represents a behavior that should never be possible, considering the topology of the house i.e. the fact that there is no door between the first bedroom and the bathroom. Consequently, an extension of the description of the zone partition and instrumentation considering the topology is proposed in the next paragraph.

## Approach 2, including the topology of the home

In this second approach (Danancher et al., 2013c), the zone partition $\mathbf{Z}$, the sensors $\mathbf{S}$ and the observed zones $Obs^+$ are still considered. In addition, the topology is considered. Topology is meant to represent the direct paths between the different zones of the house. It is called $Top$ and the following definition is given.

**Definition 10** (Topology $Top$). A *Topology* related to a Zone Partition $\mathbf{Z}$ represents the direct paths between the zones. A *Topology* is formally described by a ($|\mathbf{Z}| \times |\mathbf{Z}|$)-matrix denoted $Top$ where $Top_{(i,j)} = 1$ if there exists a direct path between zone $\mathbf{Z}_i$ and zone $\mathbf{Z}_j$, 0 else.

For the case study and the considered zone partition of Fig. 2.4, the following topology is defined. For instance, a direct path exists between the first bedroom $\mathbf{Z}_1$ and the corridor $\mathbf{Z}_2$ because $Top_{(1,2)} = 1$. On the contrary, it can be seen that there is no direct path between the first bedroom $\mathbf{Z}_1$ and the bathroom $\mathbf{Z}_4$ because $Top_{(1,4)} = 0$.

Note that $\forall i \; Top_{(i,i)} = 1$ i.e. the diagonal elements of $Top$ are all equal to 1. Moreover, the matrix $Top$ is symmetrical. Even if it is not necessary from a theoretical point of view, it is obvious considering that the doors can always be crossed in both directions.

$$Top = \begin{bmatrix} 1 & 1 & 0 & 0 & 0 & 0 & 0 & 0 \\ 1 & 1 & 1 & 0 & 1 & 1 & 1 & 0 \\ 0 & 1 & 1 & 1 & 0 & 0 & 0 & 0 \\ 0 & 0 & 1 & 1 & 0 & 0 & 0 & 0 \\ 0 & 1 & 0 & 0 & 1 & 0 & 0 & 0 \\ 0 & 1 & 0 & 0 & 0 & 1 & 0 & 0 \\ 0 & 1 & 0 & 0 & 0 & 0 & 1 & 1 \\ 0 & 0 & 0 & 0 & 0 & 0 & 1 & 1 \end{bmatrix}$$

Based on $\mathbf{Z}$, $Top$, $\mathbf{S}$ and $Obs^+$, Algorithm 2.2 is proposed to systematically generate the $DMA$.

---

**Algorithm 2.2** Generation of the $DMA$ starting from $\mathbf{Z}, Top, \mathbf{S}, Obs^+$

---

**Require:** $\mathbf{Z}, Top, \mathbf{S}, Obs^+$
1: $DMA := \langle Q, \Sigma, \delta, Q_0 \rangle$
2: **for** $i := 1$ to $|\mathbf{Z}|$ **do**
3:     Create state $q_i = \mathbf{Z}_i$ in $Q$
4:     Set state $q_i$ initial, $q_i \in Q_0$
5: **end for**
6: **for** $k := 1$ to $|\mathbf{S}|$ **do**
7:     Create event $\mathbf{S}_k\_1$ in $\Sigma$
8:     **if** $\mathbf{S}_k$ is not a motion detector **then**
9:         Create event $\mathbf{S}_k\_0$ in $\Sigma$
10:     **end if**
11: **end for**
12: **for** $(i,j,k) := (1,1,1)$ to $(|\mathbf{Z}|, |\mathbf{Z}|, |\mathbf{S}|)$ **do**
13:     **if** $(Obs^+_{(k,j)} = 1) \wedge (Top_{(i,j)} = 1)$ **then**
14:         Define $\delta(q_i, \mathbf{S}_k\_1) = q_j$
15:         **if** $\mathbf{S}_k$ is not a motion detector **then**
16:             Define $\delta(q_i, \mathbf{S}_k\_0) = q_j$
17:         **end if**
18:     **end if**
19: **end for**
20: **return** $DMA$

---

The first steps consist in exactly the same as Algorithm 2.1: the set $Q$ of states is defined (line 3), each state of the automaton is defined as being an initial state of the automaton (line 4), the events are created (lines 6 to 10). The only difference is in the definition of the transitions (lines 12 to 19). A transition exists between a state $q_i$ and a state $q_j$ labeled with the event $\mathbf{S}_k\_1$ (and a transition labeled with the event $\mathbf{S}_k\_0$ also exists for sensors not being motion detectors) if it is topologically possible ($Top_{(i,j)} = 1$) and if the zone related to the destination state $q_j$ is observed by the sensor $\mathbf{S}_k$ ($Obs^+_{(k,j)} = 1$). The notion of topologically possible refers to the direct paths defined in $Top$.

By applying this algorithm on the case study described by the previous $\mathbf{Z}$, $Top$, $\mathbf{S}$ and $Obs^+$, the $DMA$ of Fig. 2.6 is systematically built.

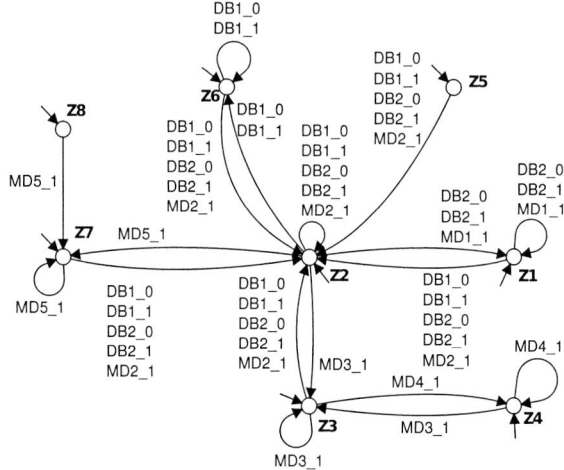

Figure 2.6.: $DMA$ generated from $\mathbf{Z}, Top, \mathbf{S}, Obs^+$ with Algorithm 2.2

This $DMA$ correctly takes into account the topology of the house, however it still contains over-reproducible behavior. For instance, due to the big size of the zones that have been chosen, it is considered that the door barrier sensor $DB_1$ is observing both the toilets $\mathbf{Z}_2$ and the corridor $\mathbf{Z}_6$ because it is placed at the interface between these two zones. Thus, by strictly applying Algorithm 2.2, there is a transition from the state $\mathbf{Z}_1$ to the state $\mathbf{Z}_2$ labeled with the events related to $DB_1$. As for the exceeding behavior of the previous $DMA$ based on approach 1, this transition has no chance to be observed under the assumptions of fault-free sensors and single inhabitant inside. Consequently, it is proposed to formalize more accurately the instrumentation using more expert knowledge in order to improve the model.

### Approach 3, using a more detailed description of the instrumentation

In this third approach, the matrix $Obs^+$ is completed with a matrix $Obs^-$. In addition, $\mathbf{Z}$, $Top$ and $\mathbf{S}$ are still considered. Using two matrices allows describing the observable change of zone in a more accurate way. For each sensor, the matrix $Obs^+$ represents the zone where the inhabitant is estimated to be after the observation of a sensor event whereas the matrix $Obs^-$ represents the zone where the inhabitant was estimated to be before the observation of a sensor event. The definition of this new matrix is given below.

**Definition 11** (Zones Before Observation $Obs^-$). The *Zones Before Observation* related to a

Zone Partition $\mathbf{Z}$ and an instrumentation $\mathbf{S}$ represent the zone(s) in which the inhabitant was before being observed by each sensor of the instrumentation. *Zones Before Observation* are formally described by a $(|\mathbf{S}| \times |\mathbf{Z}|)$-matrix denoted $Obs^-$ where $Obs^-_{(i,j)} = 1$ if it is assumed that the inhabitant was in zone $\mathbf{Z}_j$ before a motion is observed by the sensor $\mathbf{S}_i$, 0 else.

For the case study, the following matrices are proposed to represent the instrumentation. It can be seen that for the sensor $\mathbf{S}_1 = MD_1$, the inhabitant is estimated to be in $\mathbf{Z}_1$ (the first bedroom) after the observation of a sensor event $(MD_1\_1)$ and he was estimated to be in $\mathbf{Z}_1$ or in $\mathbf{Z}_2$ (the corridor) before the observation of this event because $Obs^+_{(1,1)} = 1$ and $\forall j \neq 1\ Obs^+_{(1,j)} = 0$ and $Obs^-_{(1,1)} = Obs^-_{(1,2)} = 1$ and $\forall j \notin \{1,2\}\ Obs^+_{(1,j)} = 0$. In this case, the advantage of this formalization over the usage of only $Obs^+$ is not obvious. For other cases, as door barrier sensors, it is more interesting. For instance, considering the sensor $\mathbf{S}_7 = DB_2$ between the first bedroom $\mathbf{Z}_1$ and the corridor $\mathbf{Z}_2$, the inhabitant is estimated to be in $\mathbf{Z}_1$ (the first bedroom) or in $\mathbf{Z}_2$ (the corridor) after the observation of a sensor event (rising or falling edge of the door barrier sensor) and he was estimated to be in in $\mathbf{Z}_1$ or in $\mathbf{Z}_2$ before the observation of this event because $Obs^+_{(1,1)} = Obs^+_{(1,2)} = 1$ and $\forall j \notin \{1,2\}\ Obs^+_{(1,j)} = 0$ and $Obs^-_{(1,1)} = Obs^-_{(1,2)} = 1$ and $\forall j \notin \{1,2\}\ Obs^+_{(1,j)} = 0$. The consequence on the $DMA$ will be highlighted after the presentation of the algorithm.

$$Obs^+ = \begin{bmatrix} 1 & 0 & 0 & 0 & 0 & 0 & 0 & 0 \\ 0 & 1 & 0 & 0 & 0 & 0 & 0 & 0 \\ 0 & 0 & 1 & 0 & 0 & 0 & 0 & 0 \\ 0 & 0 & 0 & 1 & 0 & 0 & 0 & 0 \\ 0 & 0 & 0 & 0 & 0 & 0 & 1 & 0 \\ 0 & 1 & 0 & 0 & 0 & 1 & 0 & 0 \\ 1 & 1 & 0 & 0 & 0 & 0 & 0 & 0 \end{bmatrix}$$

$$Obs^- = \begin{bmatrix} 1 & 1 & 0 & 0 & 0 & 0 & 0 & 0 \\ 1 & 1 & 1 & 0 & 1 & 1 & 1 & 0 \\ 0 & 1 & 1 & 1 & 0 & 0 & 0 & 0 \\ 0 & 0 & 1 & 1 & 0 & 0 & 0 & 0 \\ 0 & 1 & 0 & 0 & 0 & 0 & 1 & 1 \\ 0 & 1 & 0 & 0 & 0 & 1 & 0 & 0 \\ 1 & 1 & 0 & 0 & 0 & 0 & 0 & 0 \end{bmatrix}$$

Based on $\mathbf{Z}$, $Top$, $\mathbf{S}$, $Obs^+$ and $Obs^-$, Algorithm 2.3 is proposed to systematically generate the $DMA$.

The first steps consist in exactly the same as Algorithm 2.2: the set $Q$ of states is defined (line 3), each state of the automaton is defined as being an initial state of the automaton (line 4), the events are created (lines 6 to 10). The only difference is the definition of the transitions (lines 12 to 19). A transition exists between a state $q_i$ and a state $q_j$ labeled with the event $\mathbf{S}_k\_1$ (and a transition labeled with the event $\mathbf{S}_k\_0$ also exists for sensors not being motion detectors) if it is topologically possible ($Top_{(i,j)} = 1$) and if the zone related to the destination state $q_j$ is observed by the sensor $\mathbf{S}_k$ ($Obs^+_{(k,j)} = 1$) and if the source state represents a zone where the inhabitant was estimated to be before the sensor event related to $\mathbf{S}_k$ ($Obs^-_{(k,i)} = 1$).

By applying this algorithm on the case study described by the previous $\mathbf{Z}$, $Top$, $\mathbf{S}$, $Obs^+$ and $Obs^-$, the $DMA$ of Fig. 2.7 is systematically built.

As expected, the comparison between this $DMA$ and the previous one highlights the difference of represented motion detected by the sensor $DB_2$. Since it is installed on the door

---

**Algorithm 2.3** Generation of the $DMA$ starting from $\mathbf{Z}, Top, \mathbf{S}, Obs^+, Obs^-$

---

**Require:** $\mathbf{Z}, Top, \mathbf{S}, Obs^+, Obs^-$

  1: $DMA := \langle Q, \Sigma, \delta, Q_0 \rangle$

  2: **for** $i := 1$ to $|\mathbf{Z}|$ **do**

  3:      Create state $q_i = \mathbf{Z}_i$ in $Q$

  4:      Set state $q_i$ initial, $q_i \in Q_0$

  5: **end for**

  6: **for** $k := 1$ to $|\mathbf{S}|$ **do**

  7:      Create event $\mathbf{S}_k\_1$ in $\Sigma$

  8:      **if** $\mathbf{S}_k$ is not a motion detector **then**

  9:          Create event $\mathbf{S}_k\_0$ in $\Sigma$

10:      **end if**

11: **end for**

12: **for** $(i, j, k) := (1, 1, 1)$ to $(|\mathbf{Z}|, |\mathbf{Z}|, |\mathbf{S}|)$ **do**

13:      **if** $(Obs^+_{(k,j)} = 1) \wedge (Obs^-_{(k,i)} = 1) \wedge (Top_{(i,j)} = 1)$ **then**

14:          Define $\delta(q_i, \mathbf{S}_k\_1) = q_j$

15:          **if** $\mathbf{S}_k$ is not a motion detector **then**

16:              Define $\delta(q_i, \mathbf{S}_k\_0) = q_j$

17:          **end if**

18:      **end if**

19: **end for**

20: **return** $DMA$

---

between $\mathbf{Z}_1$ and $\mathbf{Z}_2$ the motion of the inhabitant should be represented only from these two zones to these two zones and not from all the zones topologically close to these two zones.

**Approach 4, using an even more detailed description of the topology and of the instrumentation**

Finally an approach requiring even more expert knowledge can be used. This approach is based on the description of the topology observed by the sensors using a matrix $OT$ in which each element represents a direct path between two zones being observable by a list of sensor events. This matrix combines the topology $Top$ and the instrumentation $Obs^+$ and $Obs^-$. The definition is given below.

**Definition 12** (Observable Topology $OT$). An *Observable Topology* related to a Zone Partition $\mathbf{Z}$ and an instrumentation $\mathbf{S}$ represents the direct paths between the zones observed by the possible sensor events relative to each sensor of the instrumentation. An *Observable Topology* is formally described by a $(|\mathbf{Z}| \times |\mathbf{Z}|)$-matrix denoted $OT$ where each element of this matrix is a (potentially empty) list of sensor events.

For the case study, the matrix $OT$ is the following. It can be seen that there are empty lists in the matrix. This represents either the fact that there is no direct path between the two according zones or that there is a direct path between the two zones but it is not observable by any sensor.

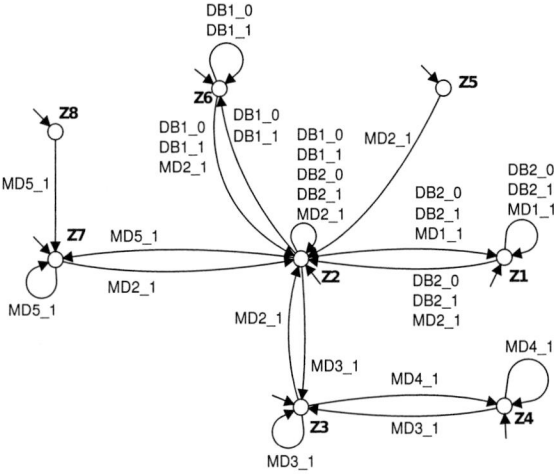

Figure 2.7.: $DMA$ generated from $\mathbf{Z}, Top, \mathbf{S}, Obs^+, Obs^-$ with Algorithm 2.3

$$
OT =
\begin{bmatrix}
\{MD_1\_1, & \{MD_2\_1, & & & & & & \\
DB_2\_1, & DB_2\_1, & \emptyset & \emptyset & \emptyset & \emptyset & \emptyset & \emptyset \\
DB_2\_0\} & DB_2\_0\} & & & & & & \\[4pt]
 & \{MD_2\_1, & & & & & & \\
\{MD_1\_1, & DB_2\_1, & & & & & & \\
DB_2\_1, & DB_2\_0, & \{MD_3\_1\} & \emptyset & \emptyset & \{DB_1\_1, & \{MD_5\_1\} & \emptyset \\
DB_2\_0\} & DB_1\_1, & & & & DB_1\_0\} & & \\
 & DB_1\_0\} & & & & & & \\[4pt]
\emptyset & \{MD_2\_1\} & \{MD_3\_1\} & \{MD_4\_1\} & \emptyset & \emptyset & \emptyset & \emptyset \\[4pt]
\emptyset & \emptyset & \{MD_3\_1\} & \{MD_4\_1\} & \emptyset & \emptyset & \emptyset & \emptyset \\[4pt]
\emptyset & \{MD_2\_1\} & \emptyset & \emptyset & \emptyset & \emptyset & \emptyset & \emptyset \\[4pt]
 & \{MD_2\_1, & & & & & & \\
\emptyset & DB_1\_1, & \emptyset & \emptyset & \emptyset & \{DB_1\_1, & \emptyset & \emptyset \\
 & DB_1\_0\} & & & & DB_1\_0\} & & \\[4pt]
\emptyset & \{MD_2\_1\} & \emptyset & \emptyset & \emptyset & \emptyset & \{MD_5\_1\} & \emptyset \\[4pt]
\emptyset & \emptyset & \emptyset & \emptyset & \emptyset & \emptyset & \{MD_5\_1\} & \emptyset
\end{bmatrix}
$$

Based on $\mathbf{Z}$, $\mathbf{S}$ and $OT$, the construction of the $DMA$ is easy to perform. As in the previous algorithms, for each $i \in \{1, ..., |\mathbf{Z}|\}$ a state $q_i = \mathbf{Z}_i$ is created in $Q$ and the state $q_i$ is initial ($q_i \in Q_0$). The set of events is constructed based on $\mathbf{S}$, for each $k \in \{1, ..., |\mathbf{S}|\}$ the event $\mathbf{S}_k\_1$ is created in $\Sigma$ and, if $\mathbf{S}_k$ is not a motion detector, the event $\mathbf{S}_k\_0$ is also created in $\Sigma$. Finally, the transitions are defined directly by using $OT$. For each $(i, j) \in \{1, ..., |\mathbf{Z}|\}^2$ there is a transition from the state $\mathbf{Z}_i$ to the state $\mathbf{Z}_j$ labeled with each event of the list $OT_{(i,j)}$. The resulting $DMA$ of this algorithm for the previously described $OT$ is exactly the same as the $DMA$ of Fig. 2.3 in this case.

Like in this particular case, this approach, in a general case, does not allow formalizing in more details the expert knowledge, it has to be seen as an alternative to the third approach.

## Discussion

Four different possible approaches are proposed in order to systematically create the model of the detectable motion ($DMA$) of a single inhabitant in his Smart Home. All these approaches are based on the same concept of partitioning the home into zones and of describing, in a more or less detailed way and requiring more or less expert knowledge, the zones observed by the different sensors. It has also been shown that the different approaches lead to more or less exceeding behavior being reproducible by the model (mainly transitions that are impossible due to the topology). Finally, in a general case the more expert knowledge is required to create the model, the less exceeding behavior is reproducible by the model. Consequently, a tradeoff between integrating exceeding behavior in the model and the level of required expert knowledge has to be found.

In the author's point of view, the second approach leads to the best tradeoff and thus this approach has been chosen to model the detectable motion of a single inhabitant.

However, the model always relies on the one hand on the quality of the description of a zone partition and instrumentation, and on the other hand on the topology of the house and on the quality of the instrumentation themselves. An approach aiming at evaluating a formalized zone partition and instrumentation (i.e. integrating the two possibilities of deviation) is presented in Chapter 4 of the present thesis.

## 2.1.5. Granularity of the model

In order to evaluate the influence of the choices of a zone partition, different possible zone partitions are considered in the following. The granularity of this choice is highlighted by considering a same home (the case study) and the same instrumentation of the home. Moreover, the same algorithm (Algorithm 2.2, considered as being the best for model building) is used to get the model.

Three examples of zone partitions are proposed, a coarse granularity where the home is split in 2 zones, a medium granularity where it is split in 10 zones and a fine granularity in 19 zones. These three zone partitions are all formalized using $\mathbf{Z}$ and $Top$. In addition, since the zones change, the matrix $Obs^+$ has to be updated regarding the new described zones. However, since the same instrumentation has been considered, the vector $\mathbf{S}$ is the same for the three different examples as previously described in the case study.

### Coarse granularity

In the case of a coarse granularity, only two zones are considered: inside and outside the house (see Fig. 2.8). The aim of performing Location Tracking with this granularity is to estimate the emptiness of the apartment. This has some applications, for instance in (Floeck, 2010), the algorithm for health problem detection based on inactivity monitoring should be deactivated when the inhabitant is outside (leading obviously to a huge inactivity in the home), in order not to raise false alarms. Another application is the automatic shutdown of dangerous devices (for instance the oven) when the house is empty.

This coarse granularity can be formalized with the following vector $\mathbf{Z}$ and matrix $Top$.

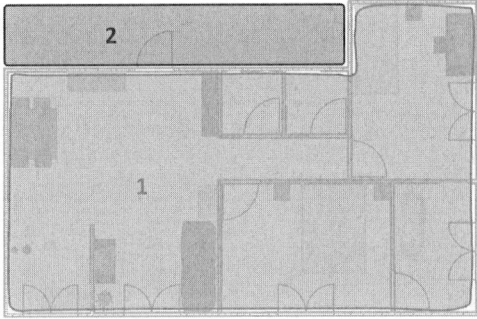

Figure 2.8.: Coarse granularity (zone partition for home occupancy monitoring)

$$\mathbf{Z} = \left[ \begin{array}{c} \mathbf{Z}_1 \\ \mathbf{Z}_2 \end{array} \right] \quad Top = \left[ \begin{array}{cc} 1 & 1 \\ 1 & 1 \end{array} \right]$$

In addition, considering the instrumentation of the case study, $Obs^+$ is defined as follows.

$$Obs^+ = \left[ \begin{array}{cc} 1 & 0 \\ 1 & 0 \\ 1 & 0 \\ 1 & 0 \\ 1 & 0 \\ 1 & 0 \\ 1 & 0 \end{array} \right]$$

Finally, by applying Algorithm 2.2, the $DMA$ of Fig. 2.9 is obtained. It can be seen in this figure that the instrumentation seems not well adapted for this zone partition since there is no transition from $\mathbf{Z}_1$ to $\mathbf{Z}_2$ i.e. it is impossible to detect the inhabitant when he is leaving the home and thus it is impossible to conclude that the house is empty.

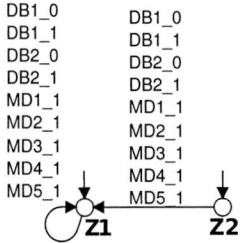

Figure 2.9.: $DMA$ for a coarse granularity

This granularity is well adapted to monitor the emptiness of the home. However, some applications are requiring a more detailed partition in zones as it is shown in the next paragraph.

**Medium granularity**

An example of medium granularity has been proposed in Fig. 2.4 where one zone is equal to one room. An alternative zone partition of the same range is proposed in Fig. 2.10 where one zone is equal to one room except the open space is divided in three zones of interest (dining room $Z_7$, kitchen $Z_8$ and living room $Z_9$).

Several applications of this granularity are relative to comfort, for instance the smart monitoring of the lights (automatically turning on the light when someone enters a zone and automatically turning off the light of the zone the inhabitant just left) or smart monitoring of the heating by learning the habits of location of the inhabitant by performing online Location Tracking. Other applications dedicated to health problem detection require this granularity, for instance applications aiming to monitor the duration of stay in the different rooms and detecting an abnormally long stay in critical rooms like the bathroom, the toilets or the shower.

Figure 2.10.: Alternative medium granularity (zone partition for duration of stay monitoring)

This alternative medium granularity can be formalized with the following vector $\mathbf{Z}$ and matrix $Top$.

$$
\mathbf{Z} = \begin{bmatrix} \mathbf{Z}_1 \\ \mathbf{Z}_2 \\ \mathbf{Z}_3 \\ \mathbf{Z}_4 \\ \mathbf{Z}_5 \\ \mathbf{Z}_6 \\ \mathbf{Z}_7 \\ \mathbf{Z}_8 \\ \mathbf{Z}_9 \\ \mathbf{Z}_{10} \end{bmatrix} \quad Top = \begin{bmatrix} 1 & 1 & 0 & 0 & 0 & 0 & 0 & 0 & 0 & 0 \\ 1 & 1 & 1 & 0 & 1 & 1 & 1 & 0 & 0 & 0 \\ 0 & 1 & 1 & 1 & 0 & 0 & 0 & 0 & 0 & 0 \\ 0 & 0 & 1 & 1 & 0 & 0 & 0 & 0 & 0 & 0 \\ 0 & 1 & 0 & 0 & 1 & 0 & 0 & 0 & 0 & 0 \\ 0 & 1 & 0 & 0 & 0 & 1 & 0 & 0 & 0 & 0 \\ 0 & 1 & 0 & 0 & 0 & 0 & 1 & 1 & 1 & 1 \\ 0 & 0 & 0 & 0 & 0 & 0 & 1 & 1 & 0 & 0 \\ 0 & 0 & 0 & 0 & 0 & 0 & 1 & 0 & 1 & 0 \\ 0 & 0 & 0 & 0 & 0 & 0 & 1 & 0 & 0 & 1 \end{bmatrix}
$$

In addition, considering the instrumentation of the case study, $Obs^+$ is defined as follows.

$$Obs^+ = \begin{bmatrix} 1 & 0 & 0 & 0 & 0 & 0 & 0 & 0 & 0 & 0 \\ 0 & 1 & 0 & 0 & 0 & 0 & 0 & 0 & 0 & 0 \\ 0 & 0 & 1 & 0 & 0 & 0 & 0 & 0 & 0 & 0 \\ 0 & 0 & 0 & 1 & 0 & 0 & 0 & 0 & 0 & 0 \\ 0 & 0 & 0 & 0 & 0 & 0 & 1 & 1 & 1 & 0 \\ 0 & 1 & 0 & 0 & 0 & 1 & 0 & 0 & 0 & 0 \\ 1 & 1 & 0 & 0 & 0 & 0 & 0 & 0 & 0 & 0 \end{bmatrix}$$

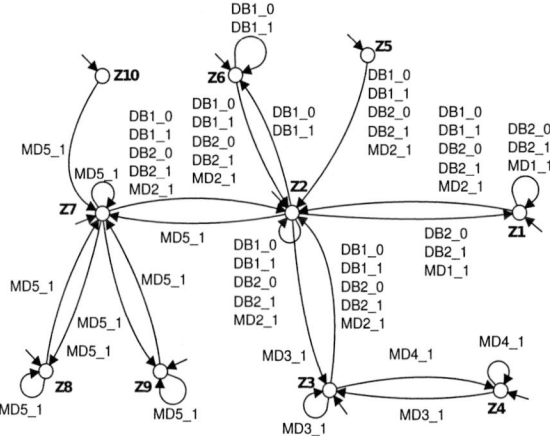

Figure 2.11.: $DMA$ for an alternative medium granularity

Finally, by applying Algorithm 2.2, the $DMA$ of Fig. 2.11 is obtained. In comparison to the $DMA$ of Fig. 2.6 (based on almost the same zone partition) it can be seen that dividing the open space in three zones is certainly interesting for some applications but with this particular instrumentation, it is impossible to determine in which one of these three zones the inhabitant is when he is observed by the sensor $MD_5$.

Still, this zone partition remains insufficiently fine for some applications (e.g. Activities of Daily Living recognition). An example of fine granularity is introduced in the next paragraph.

### Fine granularity

Applications aiming to recognize Activities of Daily Living (ADLs), and based on this recognition to suggest relevant services to the user or to prompt reminders relative to the current activity for people suffering from dementia, may be improved by knowing the location of the inhabitant. A fine granularity may help to improve the results while performing ADL recognition. This fine granularity is proposed in Fig. 2.12 where the zones are corresponding to objects of interest required to perform ADL, for instance the bathtub ($\mathbf{Z}_5$) or the sink ($\mathbf{Z}_7$) in the bathroom, the oven ($\mathbf{Z}_{15}$), the sink ($\mathbf{Z}_{17}$), the fridge ($\mathbf{Z}_{14}$) or the washing machine ($\mathbf{Z}_{16}$) in the kitchen.

This fine granularity can be formalized with the following vector $\mathbf{Z}$ and matrix $Top$.

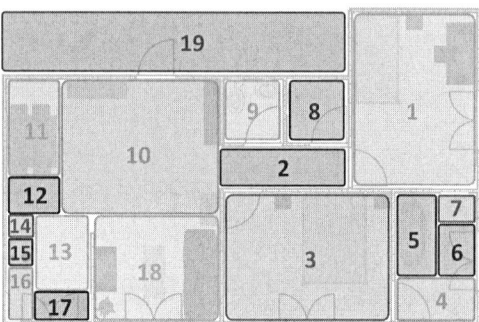

Figure 2.12.: Fine granularity (zone partition for ADL recognition)

$$
\mathbf{Z} = \begin{bmatrix} \mathbf{Z}_1 \\ \mathbf{Z}_2 \\ \mathbf{Z}_3 \\ \mathbf{Z}_4 \\ \mathbf{Z}_5 \\ \mathbf{Z}_6 \\ \mathbf{Z}_7 \\ \mathbf{Z}_8 \\ \mathbf{Z}_9 \\ \mathbf{Z}_{10} \\ \mathbf{Z}_{11} \\ \mathbf{Z}_{12} \\ \mathbf{Z}_{13} \\ \mathbf{Z}_{14} \\ \mathbf{Z}_{15} \\ \mathbf{Z}_{16} \\ \mathbf{Z}_{17} \\ \mathbf{Z}_{18} \\ \mathbf{Z}_{19} \end{bmatrix} \qquad Top = \begin{bmatrix} 1&1&0&0&0&0&0&0&0&0&0&0&0&0&0&0&0&0&0 \\ 0&1&1&0&0&0&0&1&1&1&0&0&0&0&0&0&0&0&0 \\ 0&1&1&1&0&0&0&0&0&0&0&0&0&0&0&0&0&0&0 \\ 0&0&1&1&1&1&0&0&0&0&0&0&0&0&0&0&0&0&0 \\ 0&0&0&1&1&1&1&0&0&0&0&0&0&0&0&0&0&0&0 \\ 0&0&0&1&1&1&1&0&0&0&0&0&0&0&0&0&0&0&0 \\ 0&0&0&0&1&1&1&0&0&0&0&0&0&0&0&0&0&0&0 \\ 0&1&0&0&0&0&0&1&0&0&0&0&0&0&0&0&0&0&0 \\ 0&1&0&0&0&0&0&0&1&0&0&0&0&0&0&0&0&0&0 \\ 0&1&0&0&0&0&0&0&0&1&1&1&1&0&0&0&0&1&1 \\ 0&0&0&0&0&0&0&0&0&1&1&1&0&0&0&0&0&0&0 \\ 0&0&0&0&0&0&0&0&0&1&1&1&1&1&0&0&0&0&0 \\ 0&0&0&0&0&0&0&0&0&1&0&1&1&1&1&1&1&1&0 \\ 0&0&0&0&0&0&0&0&0&0&0&1&1&1&1&0&0&0&0 \\ 0&0&0&0&0&0&0&0&0&0&0&0&1&1&1&1&0&0&0 \\ 0&0&0&0&0&0&0&0&0&0&0&0&1&0&1&1&1&0&0 \\ 0&0&0&0&0&0&0&0&0&0&0&0&1&0&0&1&1&1&0 \\ 0&0&0&0&0&0&0&0&0&1&0&0&1&0&0&0&1&1&0 \\ 0&0&0&0&0&0&0&0&0&1&0&0&0&0&0&0&0&0&1 \end{bmatrix}
$$

In addition, considering the instrumentation of the case study, $Obs^+$ is defined as follows.

$$
Obs^+ = \begin{bmatrix} 1&0&0&0&0&0&0&0&0&0&0&0&0&0&0&0&0&0&0 \\ 0&1&0&0&0&0&0&0&0&0&0&0&0&0&0&0&0&0&0 \\ 0&0&1&0&0&0&0&0&0&0&0&0&0&0&0&0&0&0&0 \\ 0&0&0&1&1&1&1&0&0&0&0&0&0&0&0&0&0&0&0 \\ 0&0&0&0&0&0&0&0&0&1&1&1&1&1&1&1&1&1&0 \\ 0&1&0&0&0&0&0&0&1&0&0&0&0&0&0&0&0&0&0 \\ 1&1&0&0&0&0&0&0&0&0&0&0&0&0&0&0&0&0&0 \end{bmatrix}
$$

Finally, by applying Algorithm 2.2, the $DMA$ of Fig. 2.13 is obtained. It obviously has many more states. However, in this particular case, the instrumentation is not adapted at all to this particular objective and it is impossible to distinguish the presence of the inhabitant among the zones 10 to 18 for instance.

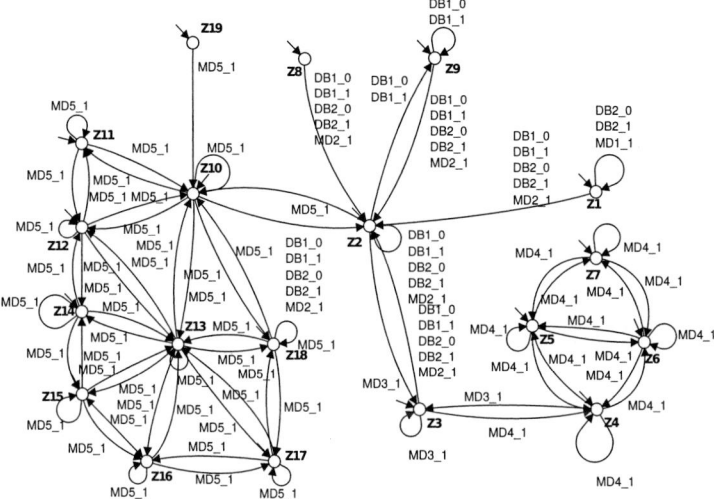

Figure 2.13.: $DMA$ for a fine granularity

One may think it is possible to make the granularity of the zone partition even finer, however there are other methods to perform Location Tracking using not only ambient binary sensors but also wearable RFID sensors for instance that will probably show better performances when a very fine granularity (under 1 meter wide zones) is required.

**Discussion**

The summary of these examples of granularity and a general conclusion on the choice that should be made for the zone partition is that it mainly depends on what the Location Tracking is aimed for.

- A coarse granularity is sufficient if the aim is to estimate the presence or absence of the inhabitant in the house. Applications are for instance the automatic shutdown of dangerous devices or the monitoring of the global inactivity in the house.

- A medium granularity (at the level of rooms or in this range) is needed if the result of the Location Tracking is aimed to be exploited by applications for smart monitoring of lights, power or heating, or for health problem detection by monitoring the duration of stay in critical zones.

- A fine granularity is required to improve the algorithms of ADL recognition with Location Tracking, thus improving the suggestion of relevant services or the prompting of reminders.

To conclude this discussion, it cannot be said that there is a zone partition being better than the others. The expert has to choose the zone partition that best fits his application of Location Tracking i.e. the zones of interest for the application. However, whatever the selected granularity for the zone partition, the approach for performance evaluation proposed in Chapter 4 can be applied.

### 2.1.6. Conclusion on the modeling for a single inhabitant

In this first section of the chapter, several approaches to model the detectable motion of a single inhabitant in a Smart Home have been presented and illustrated. No matter which approach is used the detectable motion automaton $DMA$ always satisfies Definition 6.

If the objective is to track the location of multiple inhabitants, a model of their detectable motion should be built. An approach for systematic generation of the detectable motion of $N$ inhabitants in the Smart Home is proposed in the following section. This approach takes as an input the $DMA$, respecting Definition 6.

However, if the $DMA$ is aimed to be the input for multiple inhabitants modeling, then it should contain as little behavior being reproducible by the model although being impossible in the real life (due to the topology or the instrumentation) as possible. The explanation can be illustrated on the following scenario:

Consider the case study and a $DMA$ in which it is possible to observe a direct motion from the first bedroom to the second bedroom (for instance in the $DMA$ of Fig. 2.5) which should be impossible considering the topology of the home. Consider now two inhabitants in this house, one being in the first bedroom and the second being in the second bedroom. If the first inhabitant starts moving in the first bedroom, according to the $DMA$ he is in this room. If the second inhabitant starts moving in the second room just after this first move of the first inhabitant, then it should be concluded that there are two inhabitants at home since it is impossible to go directly from the first to the second bedroom. However, if the $DMA$ contains such a transition (as the $DMA$ of Fig. 2.5), the conclusion will be that there is one inhabitant and he is now in the second bedroom. This short scenario illustrates the need for the $DMA$ to reproduce as little non-realistic behavior of a single inhabitant as possible.

However, although reducing the exceeding behavior, all the topologically possible behaviors should be included in the model. If not, there could be a loss of the Location Tracking and this would lead to bad results.

As a conclusion, the author considers that the second approach for systematic generation (based on $\mathbf{Z}$, $Top$, $\mathbf{S}$ and $Obs^+$) is a good tradeoff between required expert knowledge and quantity of exceeding behavior included in the $DMA$. Using the first systematic approach, an overly exceeding behavior is represented. The third and fourth approaches are requiring a higher level of expert knowledge even if they reduce the exceeding behavior (if the expert is competent).

## 2.2. Detectable motion of $N$ inhabitants

Based on the model of the detectable motion of a single inhabitant, an approach to model the detectable motion of multiple inhabitants is proposed (Danancher et al., 2013b). Using this approach, a model of the detectable motion of a fixed and known number of inhabitants $N$ is obtained. An overview of this approach is given in Fig. 2.14. It is composed of 3 steps where the first one is the creation of an extended model of the detectable motion of each one of the inhabitants living in the instrumented house with the $N - 1$ other inhabitants. The second step consists in performing the synchronous composition of these models in order to get a model of the detectable motion of the $N$ inhabitants being together in the house. The third step consists in reducing this model in order to have just a model of the number of inhabitants in each zone and not a model of the location of each inhabitant because it is assumed that the inhabitants are not distinguishable by the sensors. A simplified case study and these three steps are detailed in the following subsections.

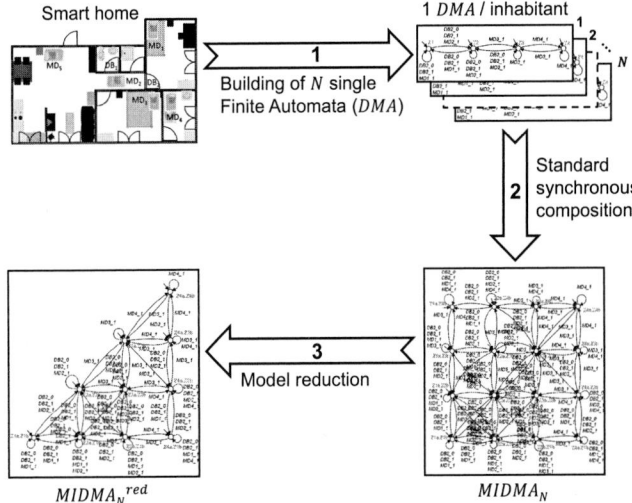

Figure 2.14.: Overview of the procedure to model the detectable motion of $N$ inhabitants

## 2.2.1. Simplified case study

Even if this approach has been successfully applied on the whole Smart Home case study of Fig. 2.3, in this section only a part of this Smart Home is considered in order for the models to be readable. Thus, only the first bedroom, the corridor, the second bedroom and the bathroom are selected as a case study. Moreover, the zone partition of Fig. 2.4 is chosen and limited to this four rooms. Consequently there are four zones and five sensors. The zone partition can be described using $\mathbf{Z}$ and $Top$ as follows:

$$\mathbf{Z} = \begin{bmatrix} \mathbf{Z}_1 \\ \mathbf{Z}_2 \\ \mathbf{Z}_3 \\ \mathbf{Z}_4 \end{bmatrix} \qquad Top = \begin{bmatrix} 1 & 1 & 0 & 0 \\ 1 & 1 & 1 & 0 \\ 0 & 1 & 1 & 1 \\ 0 & 0 & 1 & 1 \end{bmatrix}$$

In addition, considering the instrumentation of this case study, $\mathbf{S}$ and $Obs^+$ are defined as follows.

$$\mathbf{S} = \begin{bmatrix} MD_1 \\ MD_2 \\ MD_3 \\ MD_4 \\ DB_2 \end{bmatrix} \quad Obs^+ = \begin{bmatrix} 1 & 0 & 0 & 0 \\ 0 & 1 & 0 & 0 \\ 0 & 0 & 1 & 0 \\ 0 & 0 & 0 & 1 \\ 1 & 1 & 0 & 0 \end{bmatrix}$$

By applying Algorithm 2.2, the $DMA$ of Fig. 2.15 representing the detectable motion of a single inhabitant in the Smart Home is obtained.

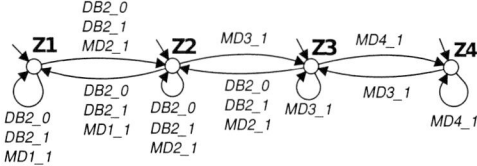

Figure 2.15.: $DMA$ for the simplified case study

## 2.2.2. Extended models of single inhabitant's detectable motion

In order to model the behavior of multiple inhabitants, it is assumed that the inhabitants behave independently one from another and thus that each one's behavior has no influence on other one's behavior. However, every inhabitant is observed by the same sensors.

Consequently, the first idea that comes to mind is to use $N$ times exactly the same $DMA$ for each inhabitant and to perform their standard synchronous composition $DMA\|DMA\| \dots \|DMA$ ($N$ times). But this would not give the expected result. Indeed, since they have the same alphabet of events $\Sigma$, the standard synchronous composition would consider them as common events.

In fact, this would not exactly represent all the possible detectable motion of the two inhabitants. Consider the following example. Two inhabitants are in the zone $\mathbf{Z}_3$. The next observed event is $MD_4\_1$ (the rising edge of the sensor observing the zone $\mathbf{Z}_4$). Since this event would be common to the two $DMA$, the result using the automaton $DMA\|DMA$ would be that the two inhabitants are now in zone $\mathbf{Z}_4$. However, the reality is more like this:

- Maybe the two inhabitants moved together to zone $\mathbf{Z}_4$

- Maybe inhabitant 1 moved to $\mathbf{Z}_4$ while inhabitant 2 stayed in $\mathbf{Z}_3$

- Maybe inhabitant 1 stayed in zone $\mathbf{Z}_3$ while inhabitant 2 moved to $\mathbf{Z}_4$.

The only certitude is that at least one person moved in zone $\mathbf{Z}_4$, assuming that the sensors have a fault free behavior.

Consequently, extended single inhabitants models are proposed. Each is aimed to represent the detectable motion of one inhabitant among the other $N-1$ inhabitants. Consider a set of inhabitants $Inh = \{Inh_1, Inh_2, ..., Inh_N\}$ and the previously created $DMA$. For each inhabitant, $DMA_{Inh_i}$ is defined for the $i^{th}$ inhabitant as $DMA_{Inh_i} = (Q_{Inh_i}, \Sigma_{Inh_i}, \delta_{Inh_i}, Q_{0_{Inh_i}})$ with:

- $Q_{Inh_i} = Q$ the set of states (associated to the zones of the house),

- $\Sigma_{Inh_i}$ an alphabet of events,

- $\delta_{Inh_i} : Q_{Inh_i} \times \Sigma_{Inh_i} \to 2^{Q_{Inh_i}}$ the transition function,

- $Q_{0_{Inh_i}} = Q_{Inh_i}$ the set of initial states.

Since the motion of each single inhabitant is represented by the previously described $DMA$, each $DMA_{Inh_i}$ has the same structure (states and transitions) as $DMA$. Each state $q_{Inh_i}$ of $Q_{Inh_i}$ has the following semantics: "inhabitant $Inh_i$ is located in zone $q$".

The motion of the inhabitants (moving between two different zones or within a zone) is observed by the sensors of the house. When observing multiple inhabitants, each time a sensor detects motion and emits an event, this event represents either one inhabitant moving alone, or one inhabitant moving at the same time as 1, 2, ..., $N-1$ other inhabitants. Different events are introduced in order to represent these different cases.

For instance, considering $N = 3$ inhabitants with $Inh = \{a, b, c\}$ and focusing on the observation of the motion of the inhabitant $a$, 4 events are defined for each observable sensor event $\sigma \in \Sigma$:

- $\sigma_a$ represents the observation of the motion of the inhabitant $a$ alone

- $\sigma_{ab}$ represents the observation of the motion of both inhabitant $a$ and inhabitant $b$

- $\sigma_{ac}$ represents the observation of the motion of both inhabitant $a$ and inhabitant $c$

- $\sigma_{abc}$ represents the observation of the motion of all the inhabitants $a$, $b$ and $c$

In this case, the alphabet of events of $DMA_a$ is therefore $\Sigma_a = \bigcup_{\sigma \in \Sigma} \{\sigma_a, \sigma_{ab}, \sigma_{ac}, \sigma_{abc}\}$ with $\Sigma$ the set of sensor events of $DMA$. In a similar manner, $\Sigma_b = \bigcup_{\sigma \in \Sigma} \{\sigma_b, \sigma_{ba}, \sigma_{bc}, \sigma_{bac}\}$ and $\Sigma_c = \bigcup_{\sigma \in \Sigma} \{\sigma_c, \sigma_{ca}, \sigma_{cb}, \sigma_{cab}\}$. This is illustrated on the case study for 3 inhabitants with $Inh = \{a, b, c\}$ in Table 2.1.

Note that $\forall (i, j) \in \{1, 2, ..., N\}^2$ the events $\sigma_{Inh_i Inh_j}$ and $\sigma_{Inh_j Inh_i}$ both represent the observation of the motion of both inhabitants $Inh_i$ and $Inh_j$ and can be merged. The same is true for the events representing the motion of $n$ inhabitants with $n \in \{2, ..., N\}$ (e.g. $\sigma_{abc} = \sigma_{bac} = \sigma_{cab}$)

Each set of events can be obtained systematically using the following procedure. Let us denote as $Part(Inh)$ the set of partitions of the set $Inh$. For $Inh = \{a, b, c\}$, $Part(Inh) = \{\emptyset, a, b, c, ab, ac, bc, abc\}$. To construct the set of events related to the inhabitant $a$, only the partitions including $a$ are considered i.e. only the partitions not being composed of only other inhabitants are considered. The set of partitions of only other inhabitants is equal to $Part(Inh \backslash \{a\}) = Part(\{b, c\}) = \{\emptyset, b, c, bc\}$. Thus, the set of only the partitions including $a$ is equal to $Part(Inh) \backslash Part(Inh \backslash \{a\}) = \{\emptyset, a, b, c, ab, ac, bc, abc\} \backslash \{\emptyset, b, c, bc\} = \{a, ab, ac, abc\}$. The alphabet of events $\Sigma_a$ can therefore be generated by duplicating the events of $\Sigma$ as follows: $\Sigma_a = \bigcup_{\sigma \in \Sigma} \left( \bigcup_{l \in Part_a} \{\sigma_l\} \right)$ with $Part_a = Part(Inh) \backslash Part(Inh \backslash \{a\})$.

This can be generalized to $N$ inhabitants with $Inh = \{Inh_1, Inh_2, ..., Inh_N\}$. For each automaton $DMA_{Ihn_i}$, the set of events $\Sigma_{Inh_i} = \bigcup_{\sigma \in \Sigma} \left( \bigcup_{l \in Part_{Inh_i}} \{\sigma_l\} \right)$ with $Part_{Inh_i} = Part(Inh) \backslash Part(Inh \backslash \{Inh_i\})$.

Table 2.1.: Alphabets of events of the $DMA$ and of the different $DMA_{Inh_i}$

| $\Sigma$ ($|\Sigma| = 6$) | $\Sigma_a$ ($|\Sigma_a| = 24$) | $\Sigma_b$ ($|\Sigma_b| = 24$) | $\Sigma_c$ ($|\Sigma_c| = 24$) |
|---|---|---|---|
| $MD_1\_1$ | $MD_1\_1_a$ | $MD_1\_1_b$ | $MD_1\_1_c$ |
| | $MD_1\_1_{ab}$ | $MD_1\_1_{ab}$ | $MD_1\_1_{ac}$ |
| | $MD_1\_1_{ac}$ | $MD_1\_1_{bc}$ | $MD_1\_1_{bc}$ |
| | $MD_1\_1_{abc}$ | $MD_1\_1_{abc}$ | $MD_1\_1_{abc}$ |
| $MD_2\_1$ | $MD_2\_1_a$ | $MD_2\_1_b$ | $MD_2\_1_c$ |
| | $MD_2\_1_{ab}$ | $MD_2\_1_{ab}$ | $MD_2\_1_{ac}$ |
| | $MD_2\_1_{ac}$ | $MD_2\_1_{bc}$ | $MD_2\_1_{bc}$ |
| | $MD_2\_1_{abc}$ | $MD_2\_1_{abc}$ | $MD_2\_1_{abc}$ |
| $MD_3\_1$ | $MD_3\_1_a$ | $MD_3\_1_b$ | $MD_3\_1_c$ |
| | $MD_3\_1_{ab}$ | $MD_3\_1_{ab}$ | $MD_3\_1_{ac}$ |
| | $MD_3\_1_{ac}$ | $MD_3\_1_{bc}$ | $MD_3\_1_{bc}$ |
| | $MD_3\_1_{abc}$ | $MD_3\_1_{abc}$ | $MD_3\_1_{abc}$ |
| $MD_4\_1$ | $MD_4\_1_a$ | $MD_4\_1_b$ | $MD_4\_1_c$ |
| | $MD_4\_1_{ab}$ | $MD_4\_1_{ab}$ | $MD_4\_1_{ac}$ |
| | $MD_4\_1_{ac}$ | $MD_4\_1_{bc}$ | $MD_4\_1_{bc}$ |
| | $MD_4\_1_{abc}$ | $MD_4\_1_{abc}$ | $MD_4\_1_{abc}$ |
| $DB_2\_1$ | $DB_2\_1_a$ | $DB_2\_1_b$ | $DB_2\_1_c$ |
| | $DB_2\_1_{ab}$ | $DB_2\_1_{ab}$ | $DB_2\_1_{ac}$ |
| | $DB_2\_1_{ac}$ | $DB_2\_1_{bc}$ | $DB_2\_1_{bc}$ |
| | $DB_2\_1_{abc}$ | $DB_2\_1_{abc}$ | $DB_2\_1_{abc}$ |
| $DB_2\_0$ | $DB_2\_0_a$ | $DB_2\_0_b$ | $DB_2\_0_c$ |
| | $DB_2\_0_{ab}$ | $DB_2\_0_{ab}$ | $DB_2\_0_{ac}$ |
| | $DB_2\_0_{ac}$ | $DB_2\_0_{bc}$ | $DB_2\_0_{bc}$ |
| | $DB_2\_0_{abc}$ | $DB_2\_0_{abc}$ | $DB_2\_0_{abc}$ |

The cardinal of the alphabet of each automaton $|\Sigma_{Inh_i}|$ is equal to the product of the cardinal of $Part_{Inh_i}$ and the number of sensor events $|\Sigma|$ with:

$$\begin{aligned} Card\big(Part_{Inh_i}\big) &= Card\big(Part(Inh) \setminus Part(Inh \setminus \{Inh_i\})\big) \\ &= Card\big(Part(Inh)\big) - Card\big(Part(Inh \setminus \{Inh_i\})\big) \\ &= 2^N - 2^{N-1} \\ &= 2^{N-1} \end{aligned}$$

Thus $|\Sigma_{Inh_i}| = |\Sigma| \times 2^{N-1}$

Finally, to obtain $DMA_{Inh_i}$, each transition defined in $DMA$ labeled by a sensor event $\sigma$ is duplicated $2^{N-1}$ times with the same source state and destination state but labeled with one of the $2^{N-1}$ events determined as described above.

For the case study with 3 inhabitants $Inh = \{a, b, c\}$, the automaton $DMA_a$ representing the motion of the inhabitant $a$ in the same house as the two other inhabitants is given in Fig. 2.16.

In this model, the events have been duplicated in order to represent the motion of one inhabitant alone or with 1, 2, ..., $N-1$ people. Nevertheless, online Location Tracking is based on a sequence of observed events. This set of sensor events is $\Sigma$, previously defined for Single Inhabitant Location Tracking.

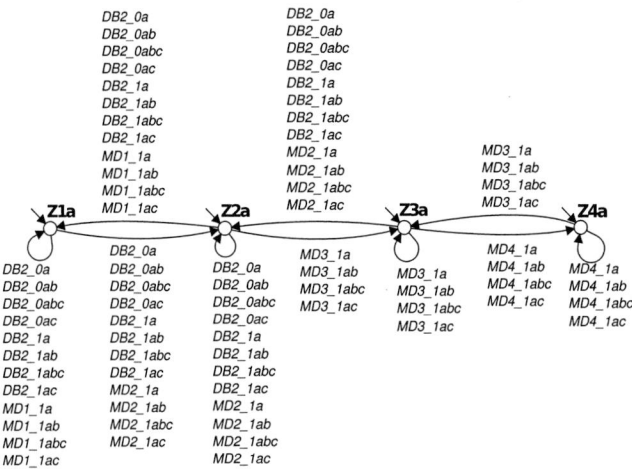

Figure 2.16.: Extended $DMA$ for the inhabitant $a$: $DMA_a$

In practice, events $DB_2\_1_{ab}$, $DB_2\_1_c$ or $DB_2\_1_{abc}$ for instance will be observed in a non-distinguishable manner by the sensor $DB_2$ through the event $DB_2\_1$.

However, this "artificial" duplication of the labeled transitions allows obtaining the finite automaton representing the detectable motion of the $N$ inhabitants together in the instrumented house by performing the standard synchronous composition as described in the following subsection.

### 2.2.3. Composition of the single models and considerations on sensor events

The definition of the standard synchronous composition of 2 finite automata is recalled below (Sampath et al., 1996) and can be generalized to $N$ automata.

**Definition 13** (Standard synchronous composition (Sampath et al., 1996)). Let $G_1$ and $G_2$ be two automata such that $G_1 = (Q_1, \Sigma_1, \delta_1, Q_{0,1})$ and $G_2 = (Q_2, \Sigma_2, \delta_2, Q_{0,2})$. The automaton $G_{sc} = (Q_{sc}, \Sigma_{sc}, \delta_{sc}, Q_{0_{sc}})$ is computed by synchronous composition of $G_1$ and $G_2$ such that: $G_{sc} = G_1 \| G_2 = (Q_{sc}, \Sigma_{sc}, \delta_{sc}, Q_{0_{sc}})$ with:

- $Q_{sc} \subseteq Q_1 \times Q_2$ ;

- $\Sigma_{sc} = \Sigma_1 \cup \Sigma_2$ ;

- $\delta_{sc}((q_1, q_2), \sigma) =$

$$
\begin{cases}
(\delta_1(q_1, \sigma), \delta_2(q_2, \sigma)) & \text{if } \delta_1(q_1, \sigma)! \text{ and } \delta_2(q_2, \sigma)! \quad (1) \\
(\delta_1(q_1, \sigma), q_2) & \text{if } \delta_1(q_1, \sigma)! \text{ and } \sigma \notin \Sigma_2 \\
(q_1, \delta_2(q_2, \sigma)) & \text{if } \delta_2(q_2, \sigma)! \text{ and } \sigma \notin \Sigma_1 \quad (2) \\
\text{undefined} & \text{otherwise}
\end{cases}
$$

- $Q_{0_{sc}} = Q_{sc} \cap (Q_{0,1} \times Q_{0,2})$

In this approach, the model of the detectable motion of multiple inhabitants is called $MIDMA_N$ (Multiple Inhabitants Detectable Motion Automaton). Its structure is given by the standard synchronous composition in the following way: $MIDAM_N = (Q_N, \Sigma_N, \delta_N, Q_{0_N}) = DMA_{Inh_1} || DMA_{Inh_2} || \ldots || DMA_{Inh_N}$.

The set of events $\Sigma_N$ is equal to the union of the sets $\Sigma_{Inh_i}$. Some events are common to two or more automata $DMA_{Inh_i}$ and some are specific to one $DMA_{Inh_i}$. The composition is performed in a synchronous manner (case (1)) on common events and in an asynchronous manner (case (2)) on non-common events. Considering the case study with 3 inhabitants and focusing on the event $MD_1\_1$:

- The three events $MD_1\_1_a$, $MD_1\_1_b$, $MD_1\_1_c$ are specific to one $DMA_{Inh_i}$, consequently, the case (2) of the definition of the transition function of the composition is applied. The transition is defined in an asynchronous manner on these non-common events.

- The three events $MD_1\_1_{ab}$, $MD_1\_1_{ac}$ and $MD_1\_1_{bc}$ are each shared by 2 automata and the event $MD_1\_1_{abc}$ belongs to all the $DMA_{Inh_i}$, thus the case (1) of the definition of the transition function of the composition is applied. The transition is defined in a synchronous manner on these common events.

After performing the composition, the resulting automaton describes all the possible motion of all inhabitants, alone or with all or part of the $N-1$ other inhabitants.

$\Sigma_N$ the set of events of $MIDMA_N$ needs now to be redefined as being equal to the set of observable sensor events $\Sigma$. For each transition of the composition labeled with an event $\sigma_N$ of $\Sigma_N$, the transition is redefined with a label being the sensor event of $\Sigma$ corresponding to $\sigma_N$. For instance, a transition labeled with the event $DB_2\_1_{ab}$ will be transformed into the same transition labeled with the event $DB_2\_1$. In the same manner, another transition labeled with the event $DB_2\_1_c$ will be transformed into the same transition labeled with the event $DB_2\_1$. Once this operation is performed, $MIDMA_N$ is usable for Multiple Inhabitants Location Tracking.

**Proposition 1.** $MIDMA_N$ has exactly $Z^N$ states, with $Z$ the number of zones and $N$ the number of inhabitants.

*Proof.* Each automaton $DMA_i$ has the same number of states $Z$ ($Z$ being the number of zones of the house). Moreover, using the approach for systematic generation of $DMA$, each state of each automaton $DMA_{Inh_i}$ is accessible with a transition labeled with one of the events $\sigma_{Inh_i}$ which is specific to the automaton i.e. $\forall (i, j) \in \{1, 2, \ldots, N\}^2$ such that $i \neq j$, $\sigma_{Inh_i} \in \Sigma_{Inh_i}$ and $\sigma_{Inh_i} \notin \Sigma_{Inh_j}$. Thus, the set of states of $DMA_{Inh_i} || DMA_{Inh_j}$ is exactly the set $Q_{Inh_i} \times Q_{Inh_j}$ whose cardinal is $Z^2$. By extension, the set of states of the composition $MIDMA_N$ of all the $DMA_{Inh_i}$ is equal to $Q_{Inh_1} \times \cdots \times Q_{Inh_N}$ and has a cardinal $Z^N$. $\square$

Note that the number of states of this synchronous composition is the same as the number of states of the asynchronous composition of the $N$ automata.

The automaton $MIDMA_2$ represents the detectable motion of $N = 2$ inhabitants with $Inh = \{a, b\}$. This automaton is given in Fig. 2.17 for the case study. Several points can be highlighted on this model.

A strong semantics is associated to the states of $MIDMA_N$. Each state represents the location of the $N$ inhabitants within the zones of the house. For instance the state $(\mathbf{Z}_{1_a}.Z_{3_b})$

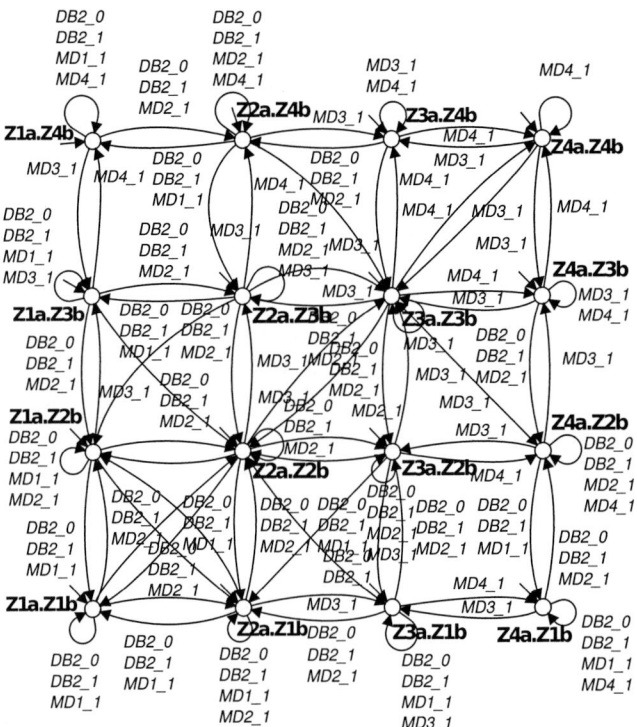

Figure 2.17.: $MIDMA_2$, composition $DMA_a||DMA_b$ for two inhabitants $\{a, b\}$, after renaming the events

means that the inhabitant $a$ is in the zone $\mathbf{Z}_1$ and the inhabitant $b$ is in the zone $\mathbf{Z}_3$. The transitions and events represent the observable motion of the inhabitants i.e. one or several inhabitants changing of location or moving in the same location in case of a self-loop.

A sensor can detect the motion of one or several inhabitants. This can be seen in the model because $MIDMA_N$ is a non-deterministic Finite Automaton (e.g. three transitions start from $(\mathbf{Z}_{4_a}.\mathbf{Z}_{4_b})$, labeled with the same event $MD_3\_1$, one reaching state $(\mathbf{Z}_{3_a}.\mathbf{Z}_{3_b})$, one reaching state $(\mathbf{Z}_{4_a}.\mathbf{Z}_{3_b})$ and one reaching $(\mathbf{Z}_{3_a}.\mathbf{Z}_{4_b})$).

It is assumed that the initial location of the inhabitant is unknown. This can be seen in the model where each state is initial. However, as explained previously for single inhabitant, knowing accurately the initial location is not necessary to perform online Location Tracking because the current estimation of the location of each inhabitant does not strongly depend on their initial location. If for some Smart Home applications it is mandatory to know the initial location of each inhabitant, some techniques (for instance in (Shu and Lin, 2013)) can be used to determine the initial state of an automaton after observing a more or less long sequence of events.

### 2.2.4. Model reduction considering non distinguishable inhabitants

Since the different inhabitants are not distinguishable by the sensors of the house, $MIDMA_N$ can be reduced using the symmetry of this model. Using a reduced model will allow performing more efficiently the online Location Tracking by having a smaller model without loss of information.

The symmetry of the model can be illustrated on the case study. If the inhabitant $a$ is in $\mathbf{Z}_1$ and the inhabitant $b$ is in $\mathbf{Z}_3$, it is exactly the same as the inhabitant $a$ is in $\mathbf{Z}_3$ and the inhabitant $b$ is in $\mathbf{Z}_1$ because the inhabitants are not distinguishable by the sensors. Consequently, the state $(\mathbf{Z}_{1_a}.\mathbf{Z}_{3_b})$ and $(\mathbf{Z}_{3_a}.\mathbf{Z}_{1_b})$ have exactly the same meaning. One of them is thus redundant and can be removed from the model. Formally for $N$ inhabitants, the state $(q_{Inh_1}.q_{Inh_2}.\cdots.q_{Inh_N})$ is equivalent to the state $(q_{Inh_2}.q_{Inh_1}.\cdots.q_{Inh_N})$ and to all the other permutations of $\{q_{Inh_1}, q_{Inh_2}, \cdots, q_{Inh_N}\}$ (e.g. $(\mathbf{Z}_{1_a}.\mathbf{Z}_{3_b}) = (\mathbf{Z}_{3_a}.\mathbf{Z}_{1_b})$ for the case study). Finally, without loss of information, the redundant states and their related transitions are deleted from the model and the obtained automaton is called $MIDMA_N^{red}$. For the case study and for two inhabitants, $MIDMA_2^{red}$ is shown in Fig. 2.18.

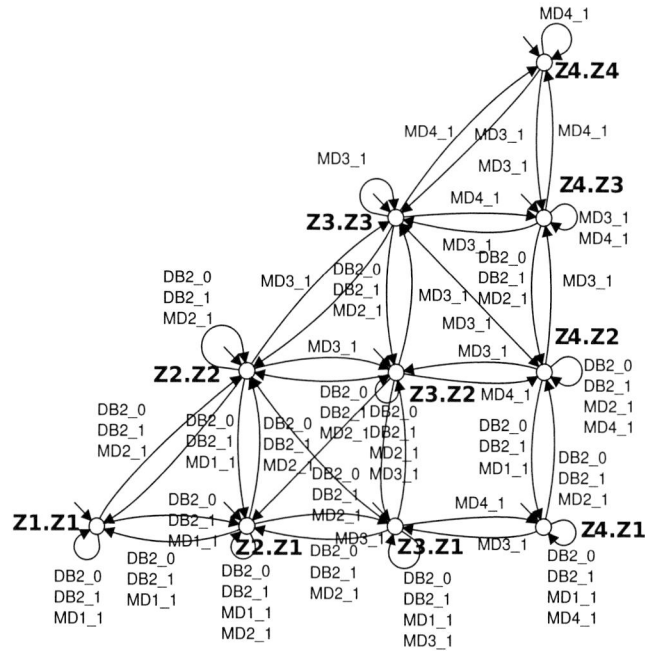

Figure 2.18.: $MIDMA_2^{red}$, reduction of $MIDMA_2$ considering non distinguishable inhabitants

The number of states of $MIDMA_N^{red}$ can be determined using the notion of multiset (see (Stanley, 2012) for more details about multisets and other enumerative problems).

**Proposition 2.** $MIDMA_N^{red}$ has exactly $\left(\!\!\left(\binom{|\mathbf{Z}|}{N}\right)\!\!\right)$ states, where $\left(\!\!\left(\binom{|\mathbf{Z}|}{N}\right)\!\!\right)$ is the number of multisets of cardinality $N$, with elements taken from a finite set of cardinality $|\mathbf{Z}|$.

*Proof.* Each state of $MIDMA_N^{red}$ represents a location of the $N$ inhabitants in the $|\mathbf{Z}|$ zones i.e. each state represents one repartition of the $N$ inhabitants among the $|\mathbf{Z}|$ zones. The number of these repartitions is equal to the number of multisets of cardinality $N$, with elements taken with repetition from a finite set of cardinality $|\mathbf{Z}|$. This number is denoted $\left(\!\!\binom{|\mathbf{Z}|}{N}\!\!\right)$ and is equal to the binomial coefficient $\binom{|\mathbf{Z}|+N-1}{N} = \dfrac{(|\mathbf{Z}| + N - 1)!}{N!(|\mathbf{Z}| - 1)!}$. □

$MIDMA_N^{red}$ represents at each time the number of inhabitants in each of the $|\mathbf{Z}|$ zones which is exactly the same information as the one represented by the automaton $MIDMA_N$ if the inhabitants are not distinguishable. Both models can be used for Location Tracking but using the reduced one leads to a decreased complexity of online Location Tracking.

For the same case study and considering 3 inhabitants, the resulting $MIDMA_3^{red}$ is shown in Fig. 2.19. A simplified version (abbreviated names of the states, partial representation of the transitions) is given in Fig. 2.20 for the sake of better understanding. Despite the number of states being not so big, the FA is hard to show. However it can be computed and then used for online Location Tracking.

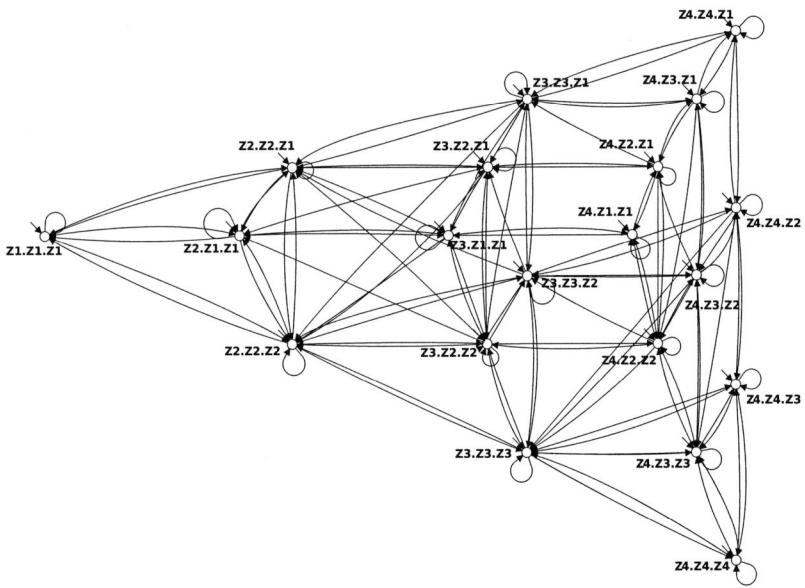

Figure 2.19.: $MIDMA_3^{red}$, model for three non distinguishable inhabitants (transition labels are not shown)

The number of states of $MIDMA_N^{red}$ is smaller than the number of states of $MIDMA_N$ which is equal to $|\mathbf{Z}|^N$ as proved in Proposition 1. $|\mathbf{Z}|^N$ is hard to compare with $\left(\!\!\binom{|\mathbf{Z}|}{N}\!\!\right)$ but a numerical illustrative comparison leads to, for $|\mathbf{Z}| = 8$ zones and $N = 4$ inhabitants, $MIDMA_N$ having 4096 states and $MIDMA_N^{red}$ having only 330 states, more than 10 times less states.

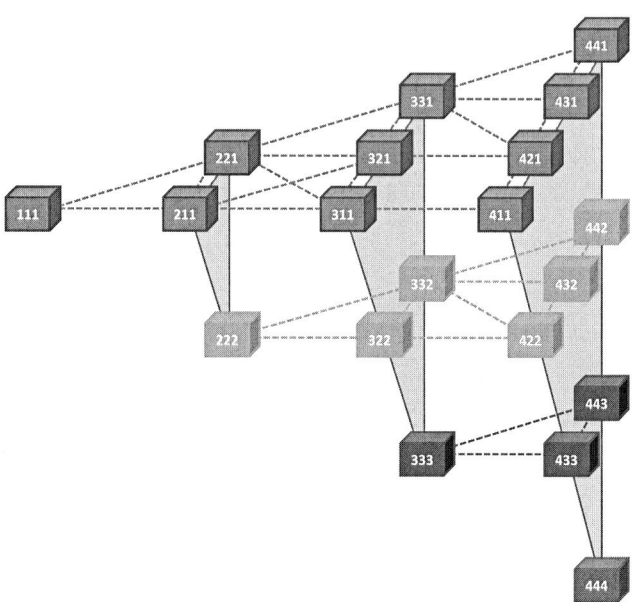

Figure 2.20.: Details of the states of $MIDMA_3^{red}$

Despite its apparent complexity, the proposed modeling approach remains scalable since we consider only instrumented apartments or houses and not a whole smart building (like for instance in (Boyer et al., 2006)). Consequently, both the number of zones $|\mathbf{Z}|$ and the number of inhabitants $N$ remain small.

### 2.2.5. Conclusion on the modeling for multiple inhabitants

In this second section of the chapter, an approach to model the detectable motion of a known number $N$ of inhabitants in a Smart Home have been presented and illustrated. This approach is based on a model of the detectable motion of a single inhabitant $DMA$ that should always satisfy Definition 6 but can be obtained by any procedure.

Obviously, other procedures could be proposed to model the detectable motion of $N$ inhabitants, for instance manual building by expert knowledge or maybe another systematic procedure, however this approach shows good performances and allows getting a model with the correct semantics of the states, events and transitions.

## Conclusion

In this chapter, a formalism to model the detectable motion of one or several inhabitants has been proposed. Several approaches to build a model for one inhabitant have been presented, applied on an illustrative case study and compared. Finally, an approach to systematically build a model of the detectable motion of $N$ inhabitants has been proposed.

This modeling procedure constitutes the first step towards model-based Location Tracking. In the next chapter, the algorithms for Location Tracking of inhabitants (a single one, a known number $N$, or an *a priori* unknown number $N$) are developed.

# 3. Online Location Tracking of the inhabitants

## Introduction

In the previous chapter, several approaches to build a model of the detectable motion of a single inhabitant or of several inhabitants in his (their) instrumented house have been presented. These models are aiming to be used for model-based online Location Tracking. An overview of this approach is given in Fig. 3.1. The inhabitants are moving in an environment instrumented with sensors; this is considered as a spontaneous event generator from the location tracker point of view. The inhabitants have a real location at each time $t$ which is called $L_{Real}(t)$. Moreover, as described before, a FA model of the detectable motion of $N$ inhabitants ($N \in \mathbb{N}^*$) is available. This model can be used by a model player (the location tracker) to estimate the current state of the FA. This estimation is called estimated location of the inhabitants $L_{Est}(t)$ and can be computed at each time $t$. $L_{Est}(t)$ is aimed to be a more or less accurate estimation of $L_{Real}(t)$ at each time.

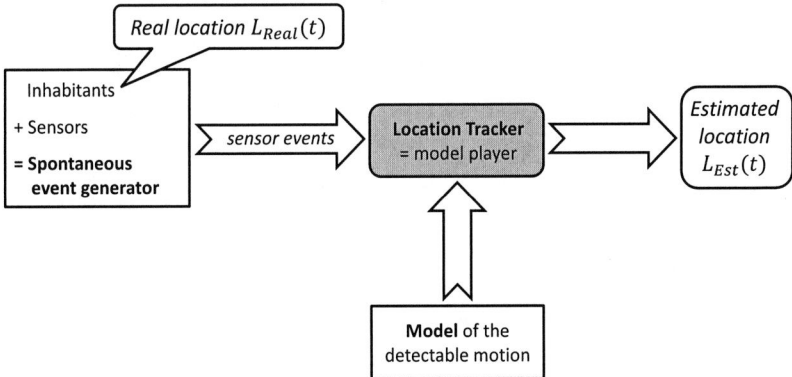

Figure 3.1.: Overview of the online model-based Location Tracking approach

In this chapter, the developed approach for online Location Tracking is presented in several steps. In a first step, named Single Inhabitant Location Tracking (SILT), it is assumed that there is always at most one inhabitant in the house(Danancher et al., 2013c). In a second step, it is considered that there are exactly $N$ inhabitants with $N \geq 2$. In a third step an algorithm for online location tracking of an *a priori* unknown number of inhabitants ($N \in \mathbb{N}$) is proposed. Finally, the possibility to relax the assumption of fault-free sensors is discussed.

## 3.1. Single Inhabitant Location Tracking

In a first time, an additional assumption is made. It is considered that there is at each time at most one inhabitant. Thus, only a model of the detectable motion of a single inhabitant, the finite automaton $DMA$, is required. Based on the $DMA$, the aim of Location Tracking is to estimate the reached state for an observed sequence of events. Since the $DMA$ is not deterministic, the construction of a state estimator of the $DMA$ is a way to perform Location Tracking. The procedure to obtain this estimator is detailed in the next subsection.

### 3.1.1. Construction of a state estimator

Since each state of $DMA$ represents the location of the inhabitant in exactly one zone of the house, the semantics of the states of the automaton is important and has to be kept while building a state estimator. This estimator is obtained by computing the equivalent deterministic automaton, by using a standard algorithm proposed in many works (Cassandras and Lafortune, 2009) or (van Glabbeek and Ploeger, 2008). Thus, the state estimator is a deterministic Finite Automaton (with a unique initial state and a deterministic transition function). Each state of the estimator is a subset of the set of states of the initial finite automaton $DMA$ and represents an estimation (more or less accurate) of the state of $DMA$. For instance, two transitions, both starting from the state $q$, both labeled with an event $\sigma$, one reaching the state $q'$ and the other the state $q''$ in the initial finite automaton $DMA$ are replaced in the estimator $Est(DMA)$ by a unique transition starting from state $q$, labeled with the event $\sigma$ and reaching the state $(q', q'')$.

The result is a FA denoted as $Est(DMA) = (Q_{Est}, \Sigma, \delta_{Est}, q_{Est_0})$ with:

- $Q_{Est} \subseteq 2^Q$ a set of states (each state is a subset of the set of states $Q$ of $DMA$),

- $\Sigma$ the same alphabet of events as the alphabet of $DMA$,

- $\delta_{Est} : Q_{Est} \times \Sigma \rightarrow Q_{Est}$ the **deterministic** transition function,

- $q_{Est_0} \in Q_{Est}$ the **unique** initial state.

The construction of a state estimator is applied to the case study of Chapter 2. The Smart Home is described in Fig. 2.3, the resulting $DMA$ is described in Fig. 2.4 and recalled in Fig. 3.2 (a). A state estimator of this $DMA$ is computed. It is denoted $Est(DMA)$ and it is shown in Fig. 3.2 (b).

As previously stated, $Est(DMA)$ has a unique initial state. Note that this state ($\mathbf{Z}_1, \mathbf{Z}_2, \mathbf{Z}_3, \mathbf{Z}_4, \mathbf{Z}_5, \mathbf{Z}_6, \mathbf{Z}_7, \mathbf{Z}_8$) represents all the states of $DMA$ since all of them are initial in $DMA$.

The state estimator $Est(DMA)$ is computed offline. This operation is theoretically of exponential complexity $O(2^{|\mathbf{Z}|})$ with $|\mathbf{Z}|$ the number of zones, but for practical applications, the number of zones is not so large and the complexity is limited.

### 3.1.2. Algorithms for model-based Location Tracking

From the state estimator, Algorithm 3.1 is proposed. Its aim is to compute the online Location Tracking that gives in real time an estimation (denoted as $L_{Est}$) of the zone(s) where the inhabitant is.

At the beginning, the estimation $L_{Est}$, representing the current location of the inhabitant, is equal to the initial state of the estimator i.e. to the set of states of the $DMA$. Then the

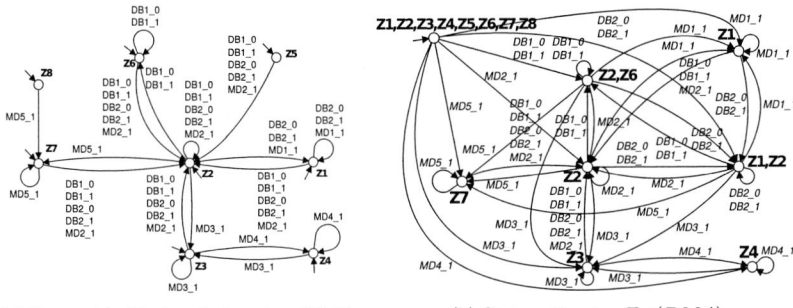

(a) Detectable Motion Automaton $DMA$      (b) State estimator $Est(DMA)$

Figure 3.2.: The detectable motion automaton $DMA$ and its state estimator $Est(DMA)$

---

**Algorithm 3.1** Estimator-based Location Tracking algorithm

---

**Require:** $Est(DMA) = (Q_{Est}, \Sigma, \delta_{Est}, q_{Est_0})$
1: Initialization of the Location Tracking:
    Current estimated location $L_{Est} = $ Initial state $q_{Est_0}$
2: **while** Location Tracking is active **do**
3:     Wait for a new event $e$
4:     New event $e$ is observed
5:     **if** $\delta_{Est}(L_{Est}, e)!$ **then**
6:         $L'_{Est} = \delta_{Est}(L_{Est}, e)$
7:         Update current estimated location $L_{Est} = L'_{Est}$
8:     **else**
9:         The estimated location remains $L_{Est}$
10:     **end if**
11: **end while**

---

algorithm waits for the occurrence of a new event $e$ emitted by a sensor. It is assumed that two events cannot simultaneously occur. When an event $e$ occurs, the algorithm computes, with respect to the transition function, the state $L'_{Est}$, successor of $L_{Est}$, such that $L'_{Est} = \delta_{Est}(L_{Est}, e)$. When the new state is computed, the state $L_{Est}$ is updated by the new state $L'_{Est}$. Then, the program waits for another new event to compute again the new location of the inhabitant.

The complexity of this algorithm is linear in the number of sensors $O(2|\mathbf{S}|)$, where $|\mathbf{S}|$ is the number of sensors. This can be demonstrated by calculating the maximal number of transitions having $L_{Est}$ as the source state that have to be explored before finding the one involving the observed event $e$. There are at most $2|\mathbf{S}|$ transitions (one involving the rising edge + one involving the falling edge for each sensor) having $L_{Est}$ as a source state. Thus, Algorithm 3.1 is efficient and the real-time computation of the online Location Tracking can be easily performed.

However, the offline computation of a state estimator is not the only option to perform online Location Tracking. The estimation of the current state of the $DMA$ can also be computed online using directly the model $DMA$. To perform online estimation of the current set of states of the $DMA$, Algorithm 3.2 is proposed.

---

**Algorithm 3.2** Online location estimation algorithm

**Require:** $DMA = (Q, \Sigma, \delta, Q_0)$

1: Initialization: Current estimated location $L_{Est}$ = set of states $Q_0$
2: **while** Location Tracking is active **do**
3:    Wait for a new event $e$
4:    New event $e$ is observed
5:    **if** $\exists q \in L_{Est}$ such that $\delta(q, e)!$ **then**
6:       $L'_{Est} = \bigcup\limits_{q \in L_{Est}} \delta(q, e)$
7:       Update current estimated location $L_{Est} = L'_{Est}$
8:    **else**
9:       The estimated location remains $L_{Est}$
10:   **end if**
11: **end while**

---

At the beginning, the estimation $L_{Est}$, representing the current location of the inhabitant, is equal to the set of the initial states of the $DMA$. Then the algorithm waits for the occurrence of the new event $e$ emitted by a sensor. It is assumed that two events cannot simultaneously occur. When an event $e$ happens, the algorithm computes, with respect to the transition function, the set of states $L'_{Est}$, successors of each state of $L_{Est}$, such that $L'_{Est} = \bigcup\limits_{q \in L_{Est}} \delta(q, e)$. When the new set of states is computed, the set of states $L_{Est}$ is updated by the new set of states $L'_{Est}$. Then, the program waits for another new event to compute again the new estimated location of the inhabitant.

The complexity of this algorithm is $O(|\mathbf{Z}| \times 2|\mathbf{S}|)$ each time a new event is observed. However, since the number of zones $|\mathbf{Z}|$ and the number of sensors $|\mathbf{S}|$ remain small in practice, this algorithm is efficient and can be performed online.

### 3.1.3. Illustration on the case study

The practical use of these algorithms can be illustrated on a real scenario of motion of the inhabitant for the case study (see Table 3.1). At each time, the Location Tracking results, either given by $Est(DMA)$ using Algorithm 3.1 or by $DMA$ using Algorithm 3.2, are the same.

The scenario is the following: the inhabitant is entering the house by the front door and is going to the first bedroom (crossing the living room, the corridor and entering shortly the shower while crossing the corridor).

0. The inhabitant is outside the house, his real location is $L_{Real} = \mathbf{Z}_8$. The Location Tracking algorithm is initialized, the estimated location $L_{Est} = (\mathbf{Z}_1, \mathbf{Z}_2, \mathbf{Z}_3, \mathbf{Z}_4, \mathbf{Z}_5, \mathbf{Z}_6, \mathbf{Z}_7, \mathbf{Z}_8)$ which is the initial state of $Est(DMA)$. It is a very ambiguous location because it is assumed the initial location of the inhabitant is unknown.

1. The inhabitant enters the house, his real location is $L_{Real} = \mathbf{Z}_7$. A rising edge of the motion detector in the living room ($MD_5\_1$) is observed. The current estimated location is updated and is now $L_{Est} = \mathbf{Z}_7$. The estimated location of the inhabitant is correct and accurate.

2. The inhabitant enters the corridor, his real location is $L_{Real} = \mathbf{Z}_2$. A rising edge of the

motion detector in the corridor ($MD_2\_1$) is observed. The current estimated location is updated and is equal to $L_{Est} = \mathbf{Z}_2$. The estimated location is still accurate.

3. The inhabitant is still in the corridor, his real location is $L_{Real} = \mathbf{Z}_2$. The falling edge of the motion detector of the living room ($MD_5\_0$) is observed. Since this event does not belong to the alphabet of events of the model $\Sigma$, the estimated location is not updated, $L_{Est} = \mathbf{Z}_2$ and the location is still accurate.

4. The inhabitant enters the shower, his real location is $L_{Real} = \mathbf{Z}_5$. No new sensor event is observed, consequently there is no update of the estimated location. $L_{Est} = \mathbf{Z}_2$ and the estimated location is incorrect.

5. The inhabitant is still in the shower, his real location is $L_{Real} = \mathbf{Z}_5$. The falling edge of the motion detector of the corridor ($MD_2\_0$) is observed. Since this event does not belong to the alphabet of events of the model $\Sigma$, the estimated location is not updated, $L_{Est} = \mathbf{Z}_2$ and the location is still incorrect.

6. The inhabitant comes back to the corridor, his real location is $L_{Real} = \mathbf{Z}_2$. A rising edge of the motion detector in the corridor ($MD_2\_1$) is observed. The current estimated location is updated and is equal to $L_{Est} = \mathbf{Z}_2$. The estimated location is accurate again.

7. The inhabitant moves to the first bedroom, his real location is still $L_{Real} = \mathbf{Z}_2$ when he starts crossing the door. A rising edge of the door barrier sensor between the corridor and the first bedroom ($DB_2\_1$) is observed. The current estimated location is updated and is equal to $L_{Est} = (\mathbf{Z}_1, \mathbf{Z}_2)$. The estimated location is ambiguous.

8. The inhabitant continues to enter the first bedroom, his real location is now $L_{Real} = \mathbf{Z}_1$. A falling edge of the door barrier sensor between the corridor and the first bedroom ($DB_2\_0$) is observed. The current estimated location is updated and is equal to $L_{Est} = (\mathbf{Z}_1, \mathbf{Z}_2)$. The estimated location is still ambiguous.

9. The inhabitant has entirely entered in the first bedroom, his real location is $L_{Real} = \mathbf{Z}_1$. A rising edge of the motion detector of the first bedroom ($MD_1\_1$) is observed. The current estimated location is updated and is equal to $L_{Est} = \mathbf{Z}_1$. The estimated location is accurate again.

The results of the Location Tracking algorithm along this scenario illustrate the three possible cases:

1. Either $L_{Real} = L_{Est}$ (step 1, 2, 3, 6, 9) and in this case, the estimated location is said to be *accurate*

2. Or $L_{Real} \in L_{Est}$ but $L_{Real} \neq L_{Est}$ (step 0, 7, 8) and in this case, the estimated location is said to be *ambiguous.*

3. Or $L_{Real} \notin L_{Est}$ (step 4, 5) and in this last case, the estimated location is said to be *incorrect.*

Obviously the first case is the most favorable for Location Tracking and the last one constitutes the worst case.

It can be seen on this small scenario that the estimated location can be incorrect only for a while and then be correct again. On the opposite, the estimation could be accurate for a while

Table 3.1.: Real scenario of motion of a single inhabitant and Single Inhabitant Location Tracking

| Step | Real location $L_{Real}$ | Observed event $e$ | Estimated location $L_{Est}$ | Estimated vs. Real |
|------|--------------------------|---------------------|------------------------------|---------------------|
| 0 | $\mathbf{Z}_8$ | $\emptyset$ | $(\mathbf{Z}_1, \mathbf{Z}_2, \mathbf{Z}_3, \mathbf{Z}_4, \mathbf{Z}_5, \mathbf{Z}_6, \mathbf{Z}_7, \mathbf{Z}_8)$ | ambiguous |
| 1 | $\mathbf{Z}_7$ | $MD_5\_1$ | $\mathbf{Z}_7$ | accurate |
| 2 | $\mathbf{Z}_2$ | $MD_2\_1$ | $\mathbf{Z}_2$ | accurate |
| 3 | $\mathbf{Z}_2$ | $MD_5\_0$ | $\mathbf{Z}_2$ | accurate |
| 4 | $\mathbf{Z}_5$ | $\emptyset$ | $\mathbf{Z}_2$ | incorrect |
| 5 | $\mathbf{Z}_5$ | $MD_2\_0$ | $\mathbf{Z}_2$ | incorrect |
| 6 | $\mathbf{Z}_2$ | $MD_2\_1$ | $\mathbf{Z}_2$ | accurate |
| 7 | $\mathbf{Z}_2$ | $DB_2\_1$ | $(\mathbf{Z}_1, \mathbf{Z}_2)$ | ambiguous |
| 8 | $\mathbf{Z}_1$ | $DB_2\_0$ | $(\mathbf{Z}_1, \mathbf{Z}_2)$ | ambiguous |
| 9 | $\mathbf{Z}_1$ | $MD_1\_1$ | $\mathbf{Z}_1$ | accurate |

and then be ambiguous again. The incorrectness or the inaccuracy of the estimation are due to the misplacement of the sensors (i.e. a zone with no installed sensors or a sensor observing several zones). Consequently, a procedure to determine *a priori* (based on the model) the possible incorrectness of Location Tracking in some zones and to evaluate *a priori* (based on the model) the ability for the location estimation to become accurate and remain accurate forever is detailed in Chapter 4.

In the next section, the assumption of a single inhabitant is relaxed and the case of $N$-Inhabitants Location Tracking ($N \geq 2$) is considered.

## 3.2. $N$-Inhabitants Location Tracking

In Chapter 2, an approach to build a model $MIDMA_N^{red}$ of the detectable motion of $N$ inhabitants in an instrumented home has been proposed. For a given number of inhabitants $N$, $N$-Inhabitants Location Tracking consists in estimating the current state of the model $MIDMA_N^{red}$, similarly to the approach for Single Inhabitant Location Tracking but with another model. Since $MIDMA_N^{red}$ is not deterministic, the construction of a state estimator is a way to perform Location Tracking. The procedure to obtain this estimator is the same as for Single Inhabitant Location Tracking; the result is given in the next subsection.

### 3.2.1. Construction of a state estimator

Using the same definition as detailed in Subsection 3.1.1, a state estimator of $MIDMA_N^{red}$ called $Est(MIDMA_N^{red})$ is computed. Considering the case study of Subsection 2.2.1 and 2 inhabitants, the $MIDMA_2^{red}$ of Fig. 2.18 is obtained. Based on this, the state estimator $Est(MIDMA_2^{red})$ is computed, it is shown in Fig. 3.3.

It is a Deterministic Finite Automaton and each of its state represents an estimation (more of less accurate) of the state of $MIDMA_N^{red}$ i.e. of the location of each of the $N$ inhabitants (the location of each of the 2 inhabitants in this case).

The state estimator $Est(MIDMA_N^{red})$ is computed offline. This operation is theoretically of exponential complexity $O(2^{|Q|})$ with $|Q| = \left(\binom{|\mathbf{Z}|}{N}\right)$ the number of states of $MIDMA_N^{red}$,

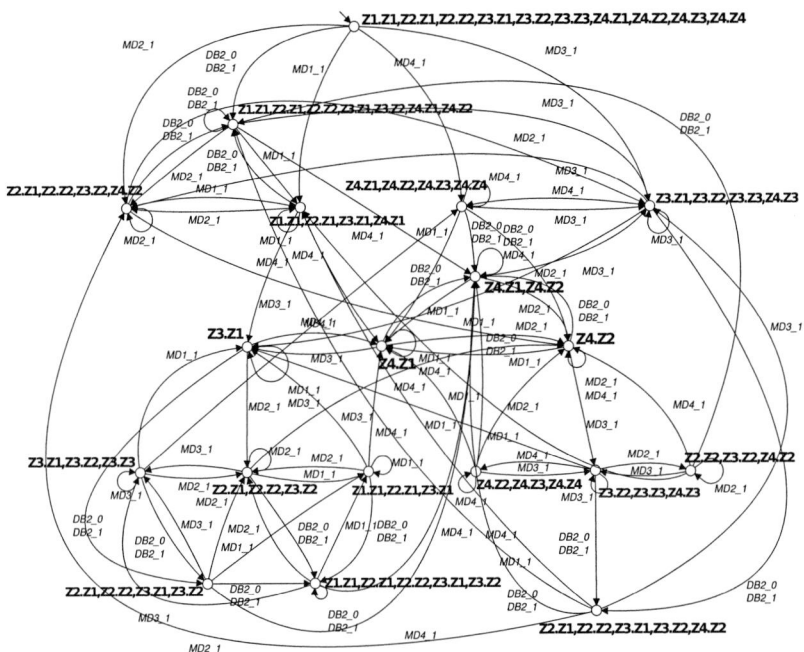

Figure 3.3.: State estimator of $MIDMA_2^{red}$: $Est(MIDMA_2^{red})$

but for practical applications, the number of zones and the number of inhabitants are not so large and the complexity is limited.

### 3.2.2. Algorithms for model-based Location Tracking

Like for Single Inhabitant Location Tracking, the online Location Tracking of $N$ inhabitants can be performed either directly on $MIDMA_N^{red}$ using Algorithm 3.2 or on the state estimator $Est(MIDMA_N^{red})$ using Algorithm 3.1. The estimated location of $N$ inhabitants is denoted as $L_{Est_N}$.

The complexity of Algorithm 3.1 is linear in the number of sensors $O(2|\mathbf{S}|)$, where $|\mathbf{S}|$ is the number of sensors. Thus, this algorithm is efficient and the real-time computation of the online Location Tracking can be easily performed.

The complexity of Algorithm 3.2 is $O\left(\binom{|\mathbf{Z}|}{N} \times 2|\mathbf{S}|\right)$ each time a new event is observed. However, since the number of zones $|\mathbf{Z}|$, the number of inhabitants $N$ and the number of sensors $|\mathbf{S}|$ remain small in practice, this algorithm is efficient and can be performed online.

### 3.2.3. Illustration on the case study

The practical use of the algorithms can be illustrated on a real scenario of motion of two inhabitants for the case study (see Table 3.2). The estimated location of 2 inhabitants is denoted as $L_{Est_2}$ while their real location is denoted as $L_{Real_2}$.

The scenario is the following: the two inhabitants are in the first bedroom, one of them moves to the corridor, enters briefly the second bedroom and comes back in the corridor. In the meantime, the other is moving sporadically inside the first bedroom.

0. The inhabitants are in the first bedroom, their real location is $L_{Real_2} = \mathbf{Z}_1.\mathbf{Z}_1$. The Location Tracking algorithm is initialized, the estimated location $L_{Est_2} = (\mathbf{Z}_1.\mathbf{Z}_1, \mathbf{Z}_2.\mathbf{Z}_1,$ $\mathbf{Z}_2.\mathbf{Z}_2, \mathbf{Z}_3.\mathbf{Z}_1, \mathbf{Z}_1.\mathbf{Z}_1, \mathbf{Z}_3.\mathbf{Z}_2, \mathbf{Z}_3.\mathbf{Z}_3, \mathbf{Z}_4.\mathbf{Z}_1, \mathbf{Z}_4.\mathbf{Z}_2, \mathbf{Z}_4.\mathbf{Z}_3, \mathbf{Z}_4.\mathbf{Z}_4)$, the initial state. It is a very ambiguous location because it is assumed the initial location of the inhabitants is unknown.

1. One inhabitant starts moving toward the corridor, the real location of the inhabitants is still $L_{Real_2} = \mathbf{Z}_1.\mathbf{Z}_1$. A rising edge of the motion detector in the first bedroom ($MD_1\_1$) is observed. The current estimated location is updated and is now $L_{Est_2} = (\mathbf{Z}_1.\mathbf{Z}_1, \mathbf{Z}_2.\mathbf{Z}_1, \mathbf{Z}_3.\mathbf{Z}_1, \mathbf{Z}_4.\mathbf{Z}_1)$. The estimated location of the inhabitants is still ambiguous, one is estimated to be in the zone $\mathbf{Z}_1$ and the other in any zone of the home $\mathbf{Z}_1, \mathbf{Z}_2, \mathbf{Z}_3$ or $\mathbf{Z}_4$.

2. This inhabitant continues moving toward the corridor, the real location of the inhabitants is still $L_{Real_2} = \mathbf{Z}_1.\mathbf{Z}_1$. A rising edge of the door barrier sensor between the first bedroom and the corridor ($DB_2\_1$) is observed. The current estimated location is updated and is now $L_{Est_2} = (\mathbf{Z}_1.\mathbf{Z}_1, \mathbf{Z}_2.\mathbf{Z}_1, \mathbf{Z}_2.\mathbf{Z}_2, \mathbf{Z}_3.\mathbf{Z}_1, \mathbf{Z}_3.\mathbf{Z}_2, \mathbf{Z}_4.\mathbf{Z}_1, \mathbf{Z}_4.\mathbf{Z}_2)$. The estimated location of the inhabitants is still ambiguous.

3. This inhabitant enters the corridor, the real location of the inhabitants is now $L_{Real_2} = \mathbf{Z}_2.\mathbf{Z}_1$. A falling edge of the door barrier sensor between the first bedroom and the corridor ($DB_2\_0$) is observed. The current estimated location is updated but is still $L_{Est_2} = (\mathbf{Z}_1.\mathbf{Z}_1, \mathbf{Z}_2.\mathbf{Z}_1, \mathbf{Z}_2.\mathbf{Z}_2, \mathbf{Z}_3.\mathbf{Z}_1, \mathbf{Z}_3.\mathbf{Z}_2, \mathbf{Z}_4.\mathbf{Z}_1, \mathbf{Z}_4.\mathbf{Z}_2)$. The estimated location of the inhabitants is still ambiguous.

4. This inhabitant continues moving in the corridor, the real location of the inhabitants is still $L_{Real_2} = \mathbf{Z}_2.\mathbf{Z}_1$. A rising edge of the motion detector of the corridor ($MD_2\_1$) is observed. The current estimated location is updated and is now $L_{Est_2} = (\mathbf{Z}_2.\mathbf{Z}_1, \mathbf{Z}_2.\mathbf{Z}_2, \mathbf{Z}_3.\mathbf{Z}_2, \mathbf{Z}_4.\mathbf{Z}_2)$. The estimated location of the inhabitants is still ambiguous.

5. This inhabitant is still in the corridor, the real location of the inhabitants is still $L_{Real_2} = \mathbf{Z}_2.\mathbf{Z}_1$. Since the second inhabitant is not moving (but still in the first bedroom) the falling edge of the motion detector of the first bedroom ($MD_1\_0$) is observed. Since this event does not belong to the alphabet of events of the model $\Sigma$, the estimated location is not updated, $L_{Est} = (\mathbf{Z}_2.\mathbf{Z}_1, \mathbf{Z}_2.\mathbf{Z}_2, \mathbf{Z}_3.\mathbf{Z}_2, \mathbf{Z}_4.\mathbf{Z}_2)$. The estimated location of the inhabitants is still ambiguous.

6. This inhabitant enters the second bedroom, the real location of the inhabitants is now $L_{Real_2} = \mathbf{Z}_3.\mathbf{Z}_1$. A rising edge of the motion detector of the second bedroom ($MD_3\_1$) is observed. The current estimated location is updated and is now $L_{Est_2} = (\mathbf{Z}_3.\mathbf{Z}_1, \mathbf{Z}_3.\mathbf{Z}_2, \mathbf{Z}_3.\mathbf{Z}_3, \mathbf{Z}_4.\mathbf{Z}_3)$. The estimated location of the inhabitants is still ambiguous.

7. In the meantime, the second inhabitant moves in the first bedroom, the real location of the inhabitants is still $L_{Real_2} = \mathbf{Z}_3.\mathbf{Z}_1$. A rising edge of the motion detector of the first bedroom ($MD_1\_1$) is observed. The current estimated location is updated and is now $L_{Est_2} = \mathbf{Z}_3.\mathbf{Z}_1$. The estimated location of the inhabitants is now accurate.

8. The inhabitants are still in their respective rooms, the real location of the inhabitants is still $L_{Real_2} = \mathbf{Z}_3.\mathbf{Z}_1$. The falling edge of the motion detector of the corridor ($MD_2\_0$) is observed. Since this event does not belong to the alphabet of events of the model $\Sigma$, the estimated location is not updated, $L_{Est} = \mathbf{Z}_3.\mathbf{Z}_1$. The estimated location of the inhabitants is still accurate.

9. Finally, the first inhabitant comes back to the corridor, the real location of the inhabitants is now $L_{Real_2} = \mathbf{Z}_3.\mathbf{Z}_2$. A rising edge of the motion detector of the corridor ($MD_2\_1$) is observed. The current estimated location is updated and is now $L_{Est_2} = (\mathbf{Z}_2.\mathbf{Z}_1, \mathbf{Z}_2.\mathbf{Z}_2, \mathbf{Z}_3.\mathbf{Z}_2)$. The estimated location of the inhabitants is ambiguous again. It cannot be decided whether the first inhabitant is back in the corridor while the second is still in the first bedroom ($\mathbf{Z}_2.\mathbf{Z}_1$) or the second inhabitant moves to the corridor while the second stays in the second bedroom ($\mathbf{Z}_3.\mathbf{Z}_2$) or if both inhabitants are entering the corridor ($\mathbf{Z}_2.\mathbf{Z}_2$).

Table 3.2.: Real scenario of motion of 2 inhabitants and 2-Inhabitants Location Tracking

| Step | Real location $L_{Real_2}$ | Observed event $e$ | Estimated location $L_{Est_2}$ | Estimated vs. Real |
|---|---|---|---|---|
| 0 | $\mathbf{Z}_1.\mathbf{Z}_1$ | $\emptyset$ | $(\mathbf{Z}_1.\mathbf{Z}_1, \mathbf{Z}_2.\mathbf{Z}_1, \mathbf{Z}_2.\mathbf{Z}_2, \mathbf{Z}_3.\mathbf{Z}_1,$ $\mathbf{Z}_3.\mathbf{Z}_2, \mathbf{Z}_3.\mathbf{Z}_3, \mathbf{Z}_4.\mathbf{Z}_1, \mathbf{Z}_4.\mathbf{Z}_2,$ $\mathbf{Z}_4.\mathbf{Z}_3, \mathbf{Z}_4.\mathbf{Z}_4)$ | ambiguous |
| 1 | $\mathbf{Z}_1.\mathbf{Z}_1$ | $MD_1\_1$ | $(\mathbf{Z}_1.\mathbf{Z}_1, \mathbf{Z}_2.\mathbf{Z}_1, \mathbf{Z}_3.\mathbf{Z}_1, \mathbf{Z}_4.\mathbf{Z}_1)$ | ambiguous |
| 2 | $\mathbf{Z}_1.\mathbf{Z}_1$ | $DB_2\_1$ | $(\mathbf{Z}_1.\mathbf{Z}_1, \mathbf{Z}_2.\mathbf{Z}_1, \mathbf{Z}_2.\mathbf{Z}_2,$ $\mathbf{Z}_3.\mathbf{Z}_1, \mathbf{Z}_3.\mathbf{Z}_2, \mathbf{Z}_4.\mathbf{Z}_1, \mathbf{Z}_4.\mathbf{Z}_2)$ | ambiguous |
| 3 | $\mathbf{Z}_2.\mathbf{Z}_1$ | $DB_2\_0$ | $(\mathbf{Z}_1.\mathbf{Z}_1, \mathbf{Z}_2.\mathbf{Z}_1, \mathbf{Z}_2.\mathbf{Z}_2,$ $\mathbf{Z}_3.\mathbf{Z}_1, \mathbf{Z}_3.\mathbf{Z}_2, \mathbf{Z}_4.\mathbf{Z}_1, \mathbf{Z}_4.\mathbf{Z}_2)$ | ambiguous |
| 4 | $\mathbf{Z}_2.\mathbf{Z}_1$ | $MD_2\_1$ | $(\mathbf{Z}_2.\mathbf{Z}_1, \mathbf{Z}_2.\mathbf{Z}_2, \mathbf{Z}_3.\mathbf{Z}_2, \mathbf{Z}_4.\mathbf{Z}_2)$ | ambiguous |
| 5 | $\mathbf{Z}_2.\mathbf{Z}_1$ | $MD_1\_0$ | $(\mathbf{Z}_2.\mathbf{Z}_1, \mathbf{Z}_2.\mathbf{Z}_2, \mathbf{Z}_3.\mathbf{Z}_2, \mathbf{Z}_4.\mathbf{Z}_2)$ | ambiguous |
| 6 | $\mathbf{Z}_3.\mathbf{Z}_1$ | $MD_3\_1$ | $(\mathbf{Z}_3.\mathbf{Z}_1, \mathbf{Z}_3.\mathbf{Z}_2, \mathbf{Z}_3.\mathbf{Z}_3, \mathbf{Z}_4.\mathbf{Z}_3)$ | ambiguous |
| 7 | $\mathbf{Z}_3.\mathbf{Z}_1$ | $MD_1\_1$ | $\mathbf{Z}_3.\mathbf{Z}_1$ | accurate |
| 8 | $\mathbf{Z}_3.\mathbf{Z}_1$ | $MD_2\_0$ | $\mathbf{Z}_3.\mathbf{Z}_1$ | accurate |
| 9 | $\mathbf{Z}_3.\mathbf{Z}_2$ | $MD_2\_1$ | $(\mathbf{Z}_2.\mathbf{Z}_1, \mathbf{Z}_2.\mathbf{Z}_2, \mathbf{Z}_3.\mathbf{Z}_2)$ | ambiguous |

Note that, contrary to the case of single inhabitant where the inaccuracy of location is only due to sensor misplacement, the inaccuracy of the estimated location of $N$ inhabitants has two sources:

- The sensors observing more than 1 zones (step 2, 3). This case corresponds exactly to the case of sensor misplacement for single inhabitant. Obviously, a sensor observing 2 zones for Single Inhabitant Location Tracking, thus leading to inaccuracy, will still observe two zones for $N$-Inhabitants Location Tracking and will also lead to inaccuracy.

- The second cause of inaccuracy comes directly from the presence of more than one inhabitant and the fact that the inhabitants are non-distinguishable (step 1, 2, 3, 4, 5, 6, 9). This is particularly illustrated in the step 9 of the scenario where either one inhabitant moves to the corridor while the other stays in the first bedroom, or the second inhabitant moves to the corridor while the other stays in the second bedroom, or both inhabitants move to the corridor. However, it is impossible to decide which one of these three possibilities is the correct one, considering only the information given by the rising edge of the sensor $MD_2$.

Like for Single Inhabitant Location Tracking, there are three possible cases for the estimated location compared to the real location (*incorrect, ambiguous, accurate*) and these situations can last more or less. Consequently, an evaluation procedure for $N$-Inhabitants Location Tracking is also proposed in Chapter 4.

The previous procedure to perform $N$-Inhabitants Location Tracking ($N \in \mathbb{N}^*$) has been developed for a predefined and constant $N$. In the following section, an approach for model-based online estimation of this number of inhabitants $N$ and Location Tracking of an *a priori* unknown number of inhabitants is proposed.

## 3.3. Location Tracking of an *a priori* unknown number of inhabitants

The considered number of inhabitants is not always known *a priori*. Moreover, it may vary depending on the date and hour of the day and on the potential guests in the house. In the following, the problem of unknown number of inhabitants Location Tracking is illustrated on the same scenario as previously used for 2-Inhabitants Location Tracking.

### 3.3.1. Illustration of the concept on the case study

The same scenario as in Subsection 3.2.3 is played. Since it is considered that the number of inhabitants is unknown, Single Inhabitant Location Tracking (SILT) and 2-Inhabitants Location Tracking (2-ILT) are performed in parallel and their results are compared. For this illustrative case study, it is considered that the maximal number of inhabitants that can be in the house is $N_{max} = 2$ and thus, there is no need to perform $N$-Inhabitants Location Tracking for $N > N_{max}$. In this case, $N_{max}$ is defined *a priori*, a procedure to estimate $N_{max}$ based on the models is detailed in Chapter 4. The results of SILT and 2-ILT for the scenario are given in Table 3.3 and explained in the following.

From step 0 to step 6, both SILT and 2-ILT give a result. The interpretation of these sequences of estimated location $L_{Est_1}$ and $L_{Est_2}$ is that the sequence of observed sensor events could have been symptomatic of the motion of 1 or 2 inhabitants. Consequently, SILT gives the result of 1 inhabitant moving (and this is mostly an accurate result) whereas the result of 2-ILT can be interpreted as 1 inhabitant is moving between the zones and in the mean time the second inhabitant is somewhere (i.e. in any zone of the house). Consider for instance the step 4, SILT gives the result $L_{Est_1} = \mathbf{Z}_2$ (the single inhabitant is in the corridor) and 2-ILT gives the result $L_{Est_2} = (\mathbf{Z}_2.\mathbf{Z}_1, \mathbf{Z}_2.\mathbf{Z}_2, \mathbf{Z}_3.\mathbf{Z}_2, \mathbf{Z}_4.\mathbf{Z}_2)$ (one inhabitant is in the corridor $\mathbf{Z}_2$ and the other is somewhere i.e. in $\mathbf{Z}_1$ or $\mathbf{Z}_2$ or $\mathbf{Z}_3$ or $\mathbf{Z}_4$).

At step 7, something interesting happens. At the end of step 6, it is considered that the single inhabitant is in the second bedroom ($L_{Est_1} = \mathbf{Z}_3$) or that one inhabitant is in the second bedroom and the other is somewhere ($L_{Est_2} = (\mathbf{Z}_3.\mathbf{Z}_1, \mathbf{Z}_3.\mathbf{Z}_2, \mathbf{Z}_3.\mathbf{Z}_3, \mathbf{Z}_4.\mathbf{Z}_3)$). Then a rising edge of the motion detector of the first bedroom ($MD_1\_1$) is observed. This event is not reproducible by the model $DMA$. Strictly applying Algorithm 3.1 on $Est(DMA)$ or Algorithm 3.2 on $DMA$ would give a result since this case in planed. In this case the

Table 3.3.: Real scenario of motion of 2 inhabitants, Single Inhabitant Location Tracking and 2-Inhabitants Location Tracking

| Step | Real location $L_{Real_2}$ | Observed event $e$ | Estimated 1-location $L_{Est_1}$ (given by SILT) | Estimated 2-location $L_{Est_2}$ (given by 2-ILT) |
|---|---|---|---|---|
| 0 | $\mathbf{Z}_1.\mathbf{Z}_1$ | $\emptyset$ | $(\mathbf{Z}_1, \mathbf{Z}_2, \mathbf{Z}_3,$ $\mathbf{Z}_4, \mathbf{Z}_5, \mathbf{Z}_6,$ $\mathbf{Z}_7, \mathbf{Z}_8)$ | $(\mathbf{Z}_1.\mathbf{Z}_1, \mathbf{Z}_2.\mathbf{Z}_1, \mathbf{Z}_2.\mathbf{Z}_2, \mathbf{Z}_3.\mathbf{Z}_1,$ $\mathbf{Z}_3.\mathbf{Z}_2, \mathbf{Z}_3.\mathbf{Z}_3, \mathbf{Z}_4.\mathbf{Z}_1, \mathbf{Z}_4.\mathbf{Z}_2,$ $\mathbf{Z}_4.\mathbf{Z}_3, \mathbf{Z}_4.\mathbf{Z}_4)$ |
| 1 | $\mathbf{Z}_1.\mathbf{Z}_1$ | $MD_1\_1$ | $\mathbf{Z}_1$ | $(\mathbf{Z}_1.\mathbf{Z}_1, \mathbf{Z}_2.\mathbf{Z}_1, \mathbf{Z}_3.\mathbf{Z}_1, \mathbf{Z}_4.\mathbf{Z}_1)$ |
| 2 | $\mathbf{Z}_1.\mathbf{Z}_1$ | $DB_2\_1$ | $(\mathbf{Z}_1, \mathbf{Z}_2)$ | $(\mathbf{Z}_1.\mathbf{Z}_1, \mathbf{Z}_2.\mathbf{Z}_1, \mathbf{Z}_2.\mathbf{Z}_2,$ $\mathbf{Z}_3.\mathbf{Z}_1, \mathbf{Z}_3.\mathbf{Z}_2, \mathbf{Z}_4.\mathbf{Z}_1, \mathbf{Z}_4.\mathbf{Z}_2)$ |
| 3 | $\mathbf{Z}_2.\mathbf{Z}_1$ | $DB_2\_0$ | $(\mathbf{Z}_1, \mathbf{Z}_2)$ | $(\mathbf{Z}_1.\mathbf{Z}_1, \mathbf{Z}_2.\mathbf{Z}_1, \mathbf{Z}_2.\mathbf{Z}_2,$ $\mathbf{Z}_3.\mathbf{Z}_1, \mathbf{Z}_3.\mathbf{Z}_2, \mathbf{Z}_4.\mathbf{Z}_1, \mathbf{Z}_4.\mathbf{Z}_2)$ |
| 4 | $\mathbf{Z}_2.\mathbf{Z}_1$ | $MD_2\_1$ | $\mathbf{Z}_2$ | $(\mathbf{Z}_2.\mathbf{Z}_1, \mathbf{Z}_2.\mathbf{Z}_2, \mathbf{Z}_3.\mathbf{Z}_2, \mathbf{Z}_4.\mathbf{Z}_2)$ |
| 5 | $\mathbf{Z}_2.\mathbf{Z}_1$ | $MD_1\_0$ | $\mathbf{Z}_2$ | $(\mathbf{Z}_2.\mathbf{Z}_1, \mathbf{Z}_2.\mathbf{Z}_2, \mathbf{Z}_3.\mathbf{Z}_2, \mathbf{Z}_4.\mathbf{Z}_2)$ |
| 6 | $\mathbf{Z}_3.\mathbf{Z}_1$ | $MD_3\_1$ | $\mathbf{Z}_3$ | $(\mathbf{Z}_3.\mathbf{Z}_1, \mathbf{Z}_3.\mathbf{Z}_2, \mathbf{Z}_3.\mathbf{Z}_3, \mathbf{Z}_4.\mathbf{Z}_3)$ |
| 7 | $\mathbf{Z}_3.\mathbf{Z}_1$ | $MD_1\_1$ | ?? | $\mathbf{Z}_3.\mathbf{Z}_1$ |
| 8 | $\mathbf{Z}_3.\mathbf{Z}_1$ | $MD_2\_0$ | ?? | $\mathbf{Z}_3.\mathbf{Z}_1$ |
| 9 | $\mathbf{Z}_3.\mathbf{Z}_2$ | $MD_2\_1$ | ?? | $(\mathbf{Z}_2.\mathbf{Z}_1, \mathbf{Z}_2.\mathbf{Z}_2, \mathbf{Z}_3.\mathbf{Z}_2)$ |

estimated location is not updated. However, an observed sensor event being non-reproducible by the model may be interpreted in two ways. A sensor fault has occurred, or there are more than one inhabitant. Since it is considered that the sensors have all a fault-free behavior, this can only be symptomatic of the presence of more than one inhabitant in the house. If there was only one inhabitant, he should have been detected in the corridor before the observation of event $MD_1\_1$. Thus, the result given by SILT should not be trusted anymore and for the step 7, 8 and 9, it is estimated that two inhabitants are present and the result of Multiple Inhabitants Location Tracking is given by $L_{Est_2}$.

Based on this illustrative scenario, an algorithm for online estimation of the current number of inhabitants and Location Tracking is proposed in the following subsection.

### 3.3.2. Algorithm for model-based number of inhabitants estimation and Location Tracking

Algorithm 3.3, aimed to estimate online the current number of inhabitants and their location, is proposed.

At the beginning, the number of inhabitants $N$ is assumed to be equal to 1. It is a conservative assumption since the case of an inhabitant being alone at home is considered as being the most critical. Indeed, health problem detection and automatic call to the emergency services are the most important when an inhabitant is alone at home. Then, Location Tracking is performed using the different models. $L_{Est_1}$ is the estimated location obtained using $Est(DMA)$, $L_{Est_2}$ is the estimated location obtained using $Est(MIDMA_2^{red})$, ..., $L_{Est_{N_{max}}}$ is the estimated location obtained using $Est(MIDMA_{N_{max}}^{red})$. Note that $N_{max}$ the maximal number of inhabitants is supposed to be known and thus that the $N_{max}$ first models have been

**Algorithm 3.3** Algorithm for online number of inhabitants estimation and Location Tracking

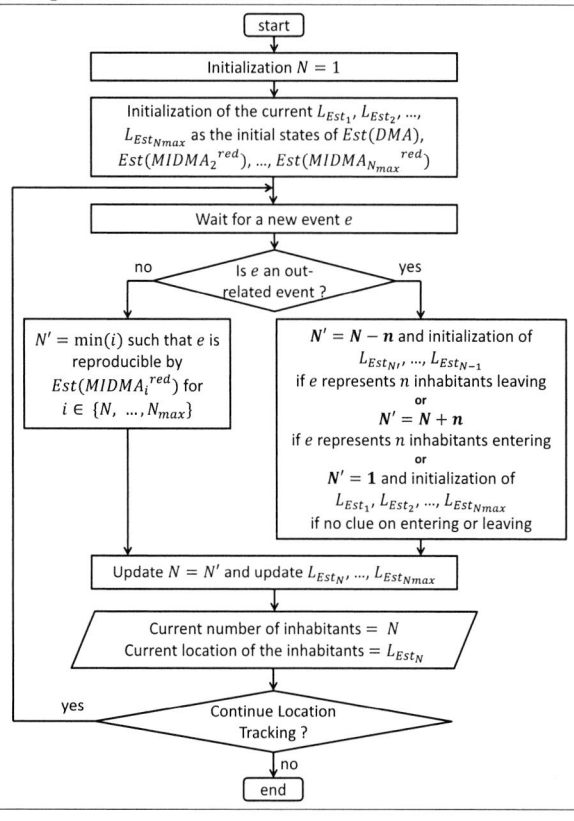

built using the different procedures detailed in Chapter 2. Moreover, a procedure to evaluate this number $N_{max}$ when it is *a priori* unknown is proposed in Chapter 4.

When a new event $e$ occurs, if the event is related to the zones being outside of the house, then it is symptomatic of a change of the current number of inhabitants in the house and thus there are three possible cases:

- The event $e$ is representative of an augmentation of $n$ of the number of inhabitants (for instance specific door barrier sensors giving the information of the direction of the inhabitant leading to $n = 1$ inhabitant entering the house), thus $N' = N + n$.

- The event $e$ is representative of a diminution of $n$ of the number of inhabitants, thus $N' = N - n$ and the models from $N'$ to $N$ should be initialized again,

- The event $e$ gives no information about one or several inhabitants entering or leaving the house, it is just symptomatic of a change of the number of inhabitants, thus $N' = 1$ in order to be back in the conservative case.

If the observed event is not related to the zones being outside of the house, the new number of inhabitants $N'$ is estimated by trying to reproduce the event $e$ with the different models $Est(MIDMA_N^{red})$, ..., $Est(MIDMA_{N_{max}}^{red})$ where $N$ is the previously determined number of inhabitants and where $Est(MIDMA_1^{red}) = Est(DMA)$. It is assumed that the sensors have all a fault-free behavior and thus, an event not being reproducible by a model $Est(MIDMA_i^{red})$ is symptomatic of the presence of strictly more than $i$ inhabitants. Consequently, the first automaton among $Est(MIDMA_N^{red})$, ..., $Est(MIDMA_{N_{max}}^{red})$ which is able to reproduce the event is the one representing the current number of inhabitants.

Once the number $N'$ is calculated (either when the event is out-related or not), $N$ is updated as being equal to $N'$ and $L_{Est_N}$, ..., $L_{Est_{N_{max}}}$ are calculated using the models and one of the Location Tracking algorithms described in the previous section (Algorithm 3.1 or Algorithm 3.2). At this time, the current number of inhabitants is estimated by $N$ and the location of these inhabitants is given by $L_{Est_N}$. Finally, the algorithm waits for a new event $e$ and starts again.

The previously proposed scenario can now be described again using Algorithm 3.3 to perform Multiple Inhabitants Location Tracking. The results are given in Table 3.4. This is quite the same result as in Table 3.3 although it is now considered that $L_{Est_1}$ has no importance from step 7 and thus is not calculated. The estimated number of inhabitants $N$ is also displayed. However, this scenario does not illustrate the case where $N$ decreases i.e. when a motion involving a door between inside the house and outside the house is observed (typically a door barrier sensor event).

Table 3.4.: Real scenario of motion of 2 inhabitants and Multiple Inhabitants Location Tracking

| Step | Real location $L_{Real_2}$ | Observed event $e$ | Estimated 1-location $L_{Est_1}$ (given by SILT) | Estimated 2-location $L_{Est_2}$ (given by 2-ILT) | Estimated number of inh. $N$ |
|---|---|---|---|---|---|
| 0 | $\mathbf{Z}_1.\mathbf{Z}_1$ | $\emptyset$ | $(\mathbf{Z}_1, \mathbf{Z}_2, \mathbf{Z}_3,$ $\mathbf{Z}_4, \mathbf{Z}_5, \mathbf{Z}_6,$ $\mathbf{Z}_7, \mathbf{Z}_8)$ | $(\mathbf{Z}_1.\mathbf{Z}_1, \mathbf{Z}_2.\mathbf{Z}_1, \mathbf{Z}_2.\mathbf{Z}_2, \mathbf{Z}_3.\mathbf{Z}_1,$ $\mathbf{Z}_3.\mathbf{Z}_2, \mathbf{Z}_3.\mathbf{Z}_3, \mathbf{Z}_4.\mathbf{Z}_1, \mathbf{Z}_4.\mathbf{Z}_2,$ $\mathbf{Z}_4.\mathbf{Z}_3, \mathbf{Z}_4.\mathbf{Z}_4)$ | 1 |
| 1 | $\mathbf{Z}_1.\mathbf{Z}_1$ | $MD_1\_1$ | $\mathbf{Z}_1$ | $(\mathbf{Z}_1.\mathbf{Z}_1, \mathbf{Z}_2.\mathbf{Z}_1, \mathbf{Z}_3.\mathbf{Z}_1, \mathbf{Z}_4.\mathbf{Z}_1)$ | 1 |
| 2 | $\mathbf{Z}_1.\mathbf{Z}_1$ | $DB_2\_1$ | $(\mathbf{Z}_1, \mathbf{Z}_2)$ | $(\mathbf{Z}_1.\mathbf{Z}_1, \mathbf{Z}_2.\mathbf{Z}_1, \mathbf{Z}_2.\mathbf{Z}_2,$ $\mathbf{Z}_3.\mathbf{Z}_1, \mathbf{Z}_3.\mathbf{Z}_2, \mathbf{Z}_4.\mathbf{Z}_1, \mathbf{Z}_4.\mathbf{Z}_2)$ | 1 |
| 3 | $\mathbf{Z}_2.\mathbf{Z}_1$ | $DB_2\_0$ | $(\mathbf{Z}_1, \mathbf{Z}_2)$ | $(\mathbf{Z}_1.\mathbf{Z}_1, \mathbf{Z}_2.\mathbf{Z}_1, \mathbf{Z}_2.\mathbf{Z}_2,$ $\mathbf{Z}_3.\mathbf{Z}_1, \mathbf{Z}_3.\mathbf{Z}_2, \mathbf{Z}_4.\mathbf{Z}_1, \mathbf{Z}_4.\mathbf{Z}_2)$ | 1 |
| 4 | $\mathbf{Z}_2.\mathbf{Z}_1$ | $MD_2\_1$ | $\mathbf{Z}_2$ | $(\mathbf{Z}_2.\mathbf{Z}_1, \mathbf{Z}_2.\mathbf{Z}_2, \mathbf{Z}_3.\mathbf{Z}_2, \mathbf{Z}_4.\mathbf{Z}_2)$ | 1 |
| 5 | $\mathbf{Z}_2.\mathbf{Z}_1$ | $MD_1\_0$ | $\mathbf{Z}_2$ | $(\mathbf{Z}_2.\mathbf{Z}_1, \mathbf{Z}_2.\mathbf{Z}_2, \mathbf{Z}_3.\mathbf{Z}_2, \mathbf{Z}_4.\mathbf{Z}_2)$ | 1 |
| 6 | $\mathbf{Z}_3.\mathbf{Z}_1$ | $MD_3\_1$ | $\mathbf{Z}_3$ | $(\mathbf{Z}_3.\mathbf{Z}_1, \mathbf{Z}_3.\mathbf{Z}_2, \mathbf{Z}_3.\mathbf{Z}_3, \mathbf{Z}_4.\mathbf{Z}_3)$ | 1 |
| 7 | $\mathbf{Z}_3.\mathbf{Z}_1$ | $MD_1\_1$ | - | $\mathbf{Z}_3.\mathbf{Z}_1$ | 2 |
| 8 | $\mathbf{Z}_3.\mathbf{Z}_1$ | $MD_2\_0$ | - | $\mathbf{Z}_3.\mathbf{Z}_1$ | 2 |
| 9 | $\mathbf{Z}_3.\mathbf{Z}_2$ | $MD_2\_1$ | - | $(\mathbf{Z}_2.\mathbf{Z}_1, \mathbf{Z}_2.\mathbf{Z}_2, \mathbf{Z}_3.\mathbf{Z}_2)$ | 2 |

Note that an important assumption regarding the fault-free sensors has been made in order for the Location Tracking of multiple inhabitants to perform well. In the following section, the possibility to relax this assumption is discussed.

## 3.4. Influence of sensor faults on the Location Tracking procedure

In this section (and in this section only), the assumption of fault-free sensors is relaxed in order to discuss the influence of sensor faults on Location Tracking results. In a first subsection, general considerations and definitions relative to faults are given. In the second subsection, several existing approaches for model-based Fault Detection and Isolation in DES are introduced and their application for Smart Home sensors FDI is discussed. Finally, the possibility of performing fault tolerant Location Tracking is discussed in the last subsection.

### 3.4.1. General considerations on faults

The definition of a *fault* needs to be clarified. Definition 14 is given in (ANSI/IEEE100, 1997).

**Definition 14** (Fault (ANSI/IEEE100, 1997)). A *fault* is a physical condition that causes a device, a component, or an element to fail to perform in a required manner, for example, a short-circuit, a broken wire, an intermittent connection.

The two terms *fault* and *failure* are often used to describe a malfunction of a component. Definition 15 highlights the difference between those two terms.

**Definition 15** (Failure (ANSI/IEEE100, 1997)). A *failure* is the termination of the ability of an item to perform in a required function.

Whereas the *failure* leads to the termination of the ability of a component to perform well, a *fault* may just be intermittent or it could be permanent but not directly influencing on the ability of the component to perform well.

Thus, it is considered that a fault may occur in a component without leading to the failure of this component. Moreover and to precise another term, one or several components of a system may have a failure without leading to the *breakdown* of the entire system. To summarize this, a component may have a *fault*. This *fault* may lead to the *failure* of the component. This *failure* may lead to the *breakdown* of the whole system.

A classification of the possible faults for industrial systems has been proposed in (Danancher et al., 2011). Considering a manufacturing system as a closed-loop DES involving the controller (typically a Programmable Logic Controller PLC), the actuators, the plant and finally the sensors, the faults can be classified in four categories:

- *Controller faults* are faults involving the controller (for instance, the computing time being longer as usual or a usually cyclic computing being not cyclic)

- *Actuator faults* are faults involving either an actuator (for instance a cylinder or a motor), a pre-actuator (for instance a single or double solenoid valve), a connection (linking an actuator to its pre-actuator or a pre-actuator to the controller) or the output unit of the controller.

- *Sensor faults* are faults involving either a sensor (for instance a position sensor), a connection (linking a sensor to the controller) or the input unit of the controller.

- *Process faults* are faults involving the process in the plant (for instance a parcel falling down a conveyor or an operator dropping off an additional parcel on a conveyor).

As in (Danancher et al., 2011) it is supposed that the controller has a fault-free behavior. Moreover, for the particular case of Location Tracking in Smart Homes, actuators are not considered. Thus, there are no possible actuator faults. Contrary to a manufacturing system, there is no plant and thus no considered process faults since the Location Tracking is aimed to be robust to any behavior of the inhabitant. Finally, only sensor faults are taken into account.

Considering the particular case of home automation, some sensor faults have been summarized in (Floeck, 2010). They are assumed to be symptomatic of the faults that can occur in Smart Homes, impacting the performance of Location Tracking. Some of these typical faults are detailed below:

- Spurious signals of motion detectors. Some of the installed motion detectors fire at night for no apparent reason. This rarely occurs.

- Motion detector's faulty activation by sunlight. Some particular sensors may be triggered due to their placement and a particular sunlight.

- Switch-off delays of motion detectors. Although the rising edge of a motion detector is observed directly after the motion of an inhabitant is effective, the falling edge is not observed directly after the inhabitant stops moving or goes out of the area observed by the sensor. This delay can be quite long (minimum 12 second in (Floeck, 2010)) and its standard variation around the average value is high. This may lead to a malfunction of the Location Tracking algorithm since someone moving alternatively between the rooms quickly may not be detected anymore when entering again a room he just left because the motion detector would not have the time to have a falling edge again.

- Power supply of sensors. In the particular case of (Floeck, 2010), some sensors are solar-powered. Consequently, some of them may suffer from a lack of sunlight and no more signals would be observed. This may be an intermittent fault since the sensors may be powered again later.

Sensor faults in Smart Homes are also considered in (Rahal et al., 2008). According to these authors, sensors may sometimes send false information. This may happen because of an intrinsic error, which can be due to the sensor error rate or to an occasional error in the experimental setup as a whole. External factors can also cause false sensor information. For instance, a draft can close a door and thus trigger a false event.

All these reviewed sensor faults are not aimed to be exhaustive. In real implemented Smart Homes or living labs, intermittent or permanent sensor faults may occur. Thus, Smart Home approaches need to be robust to the occurrence of these faults. Particularly, the Location Tracking presented in the previous section may be impacted in a wrong manner by sensor faults. The algorithm may conclude to the presence of more than one person in the house if a sensor fault occurs at a certain time. Obviously, the occurrence of a sensor fault may also lead to a wrong estimation of the location of the inhabitant(s). Consequently, the implementation of an FDI approach inspired from those dedicated to industrial systems is proposed in the next section. The ultimate aim is to be able to handle the occurrence of any sensor faults (absence of emission of an expected signal, spurious emission of an unexpected signal, ...) and to perform fault tolerant Location Tracking.

## 3.4.2. Fault Detection and Isolation

Several approaches were developed to perform FDI (also called Fault Diagnosis) for DES and particularly industrial systems. Most of them are either data-driven or model-based approaches for FDI. Since a model has been proposed in the previous chapter, only model-based FDI approaches are considered in the following.

In (Danancher et al., 2011), three of these model-based approaches were selected and compared while being applied on a same case study. Based on the results of this comparative study, the same three approaches (the diagnoser approach, the templates approach and the residual approach) have been envisaged for Smart Home sensor FDI. Their potential application for Smart Home FDI is discussed in the following paragraphs.

In (Sampath et al., 1996), a FDI approach using a so called *diagnoser* model is proposed. The diagnoser is a model of both the fault-free behavior and the faulty behavior of the considered system. One of the advantages of this method is the guarantee to detect the modeled faults (diagnosability). This advantage hides a drawback which is the non-guarantee to detect non-modeled faults. Thus, all the potential faults of the sensors, actuators or plant (in the case of a manufacturing system) should be envisaged and modeled *a priori*. Moreover, only permanent faults are considered in this method and faults involving spurious emission of unexpected signals are difficult to model. This makes this approach not adapted to the problem of Smart Home sensor faults. Furthermore, the construction of the diagnoser is based on models of the process and of each sensor or actuator (including their potential faults as unobservable events). A model of the process in the case of Smart Homes could be the $DMA$. Models of sensors could also be built by experts. However, these models are composed using the standard synchronous composition and then, a sensor map defining the value of each sensor for each state of the composition should be defined by hand. This task becomes really hard to do even with a limited number of sensors because the number of considered faults (the aim is to model the maximal number of possible faults) would drastically increase the size of the model. All these considerations make a Diagnoser-based FDI approach and, in a general way, all the approaches based on modeled faults, not-adapted for the Smart Home application.

Thus, it is believed that the best way to take into account all the faults is to consider only a model of the fault-free behavior and to monitor the deviation between the observed behavior and the modeled and expected behavior. In (Pandalai and Holloway, 2000) a FDI approach called the Templates-based FDI approach is described. This approach is based on elementary models of the timed fault-free behavior of each sensor of an industrial system. Thus, templates of the behavior of these sensors should be built by an expert and then are monitored online in order to perform the FDI. However, this approach is not adapted to the case of Smart Home either, because the behavior of the inhabitants is assumed to be totally free. The different motions or actions of each inhabitant may last more or less long, depending on his own will. Consequently, it seems impossible to propose a reliable fault-free model of the delay between occurrences of sensor events for Smart Homes. These timed approaches are not applicable for the Smart Home application.

Another approach based on a model of the fault-free behavior is the residual-based FDI, presented in (Roth et al., 2009) and illustrated in Fig. 3.4. Based on a Finite Automaton (FA) model of the *acceptable behavior* of the considered closed-loop DES, the input and output values of the controller of the DES are read online by an external FDI tool. This sequence of actuator and sensor events is called the *observed behavior* of the DES. If this *observed behavior* is reproducible by the model of the *acceptable behavior* (i.e. if there is no deviation between the *observed* and the *acceptable behavior*), then it can be concluded that the DES is functioning in an acceptable manner.

Note that, as previously described, the best case is to consider a *fault-free behavior* instead of an *acceptable behavior*. However, the guarantee that the model is representing a *fault-free behavior* is hard to give and thus, the term *acceptable behavior* is preferred. The first part of FDI, i.e. Fault Detection, is performed by detecting a deviation between the *observed behavior* and the *acceptable behavior*.

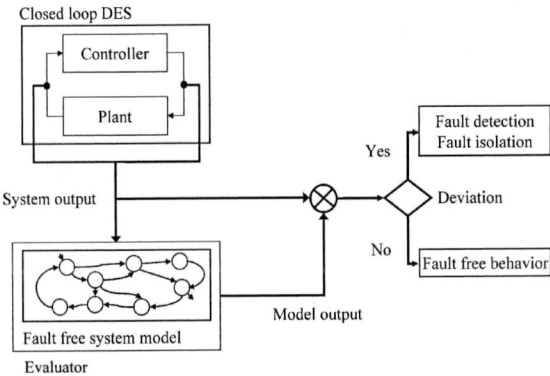

Figure 3.4.: Overview of the residual-based FDI approach (Roth, 2010)

In a second step, once a fault has been detected, the residuals can be computed in order to perform Fault Isolation (Roth, 2010; Roth et al., 2009). Computing the residuals gives a set of candidates (sensors and actuators) that are aimed to be involved in the detected fault.

The Residual-based FDI approach is strongly relying on the model of the *acceptable behavior*. Several methods can be envisaged to obtain this FA model. It can be done by expert knowledge for instance. In (Roth, 2010), an approach to identify a FA model of the *acceptable behavior* based on observed sequences of acceptable functioning of the DES is proposed. During this learning phase, the sequences have to be labeled "acceptable" by an expert. This identification approach is not adapted to the particular Smart Home issue because strongly relying on the observed behavior during the learning phase i.e. strongly depending on the considered inhabitant. Moreover, a model of the detectable motion of 1, 2, ..., $N$ inhabitants has been proposed in Chapter 2. This model can easily be judged as being a model of the *acceptable behavior* and thus can be used for the residual-based FDI approach.

An adaptation of the Residual-based FDI approach is proposed in order to deal with sensor faults in Smart Homes. As written previously, the whole approach relies on the occurrence of non-reproducible behavior. But in Algorithm 3.3, an observed behavior being non reproducible by the model $MIDMA_N^{red}$ is symptomatic of the presence of more than $N$ inhabitants (under the assumption of fault-free sensors). Yet, an observed behavior being non reproducible by the model $MIDMA_N^{red}$ is symptomatic of the presence of more than $N$ or of the occurrence of a sensor fault. Consequently, to be able to detect sensor faults, the presence of a maximal number of inhabitants should be guaranteed. This guarantee should be considered as an additional assumption, required to perform FDI. This number is called $N_{maxFDI}$.

An overview of the proposed approach is given in Fig. 3.5. The inhabitants in the Smart

Home are still considered as a spontaneous event generator from the location tracker point of view. It is assumed that the presence of less than $N_{maxFDI}$ inhabitants in the Smart Home is guaranteed. The models of the detectable motion could be obtained for 1, 2, ..., $N_{maxFDI}$ inhabitants using the algorithms of Chapter 2. The location tracker is completed with a FDI algorithm. The whole algorithm for online Location Tracking and sensor FDI is given in Algorithm 3.4. It is of course based on the algorithm for Multiple Inhabitants Location Tracking (Algorithm 3.3).

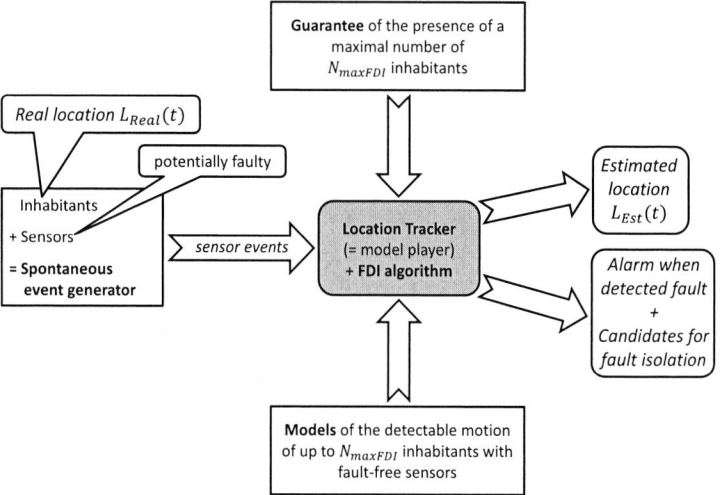

Figure 3.5.: Overview of the online model-based Location Tracking and residual-based sensor FDI approach

The practical use of this algorithm is illustrated on a real scenario in the next section.

### 3.4.3. Illustration on a case study

To illustrate the functioning of this algorithm, the scenario of motion of one inhabitant (exactly the same scenario as in Subsection 3.1.3) is envisaged. It is considered that there is at most 1 inhabitant ($N_{maxFDI} = 1$). The scenario is the following: the inhabitant is entering the house by the front door and is going to the first bedroom (crossing the living room, the corridor and entering shortly the shower while crossing the corridor). Different configurations for the sensors are considered. The case of fault free sensors is treated in Subsection 3.1.3. A first considered configuration is the spurious activation of the sensor $MD_1$ (because of the sun for instance), while the inhabitant is still in the living room. A second considered configuration is the failure of the power-supply of the sensor $DB_2$ at the very beginning of the scenario, the sensor is stuck at 0. A third considered configuration is the failure of the sensor $MD_2$ at the very beginning of the scenario, the sensor is stuck at 0. These three illustrative cases are aimed to show the impact of sensor faults on the result of Location Tracking and to show the result of the Residual-based FDI approach.

Moreover, it is assumed that the maximal number of inhabitants $N_{maxFDI} = 1$ is guaranteed.

**Algorithm 3.4** Algorithm for online number of inhabitants estimation, Location Tracking and sensor FDI

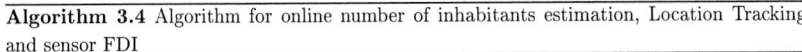

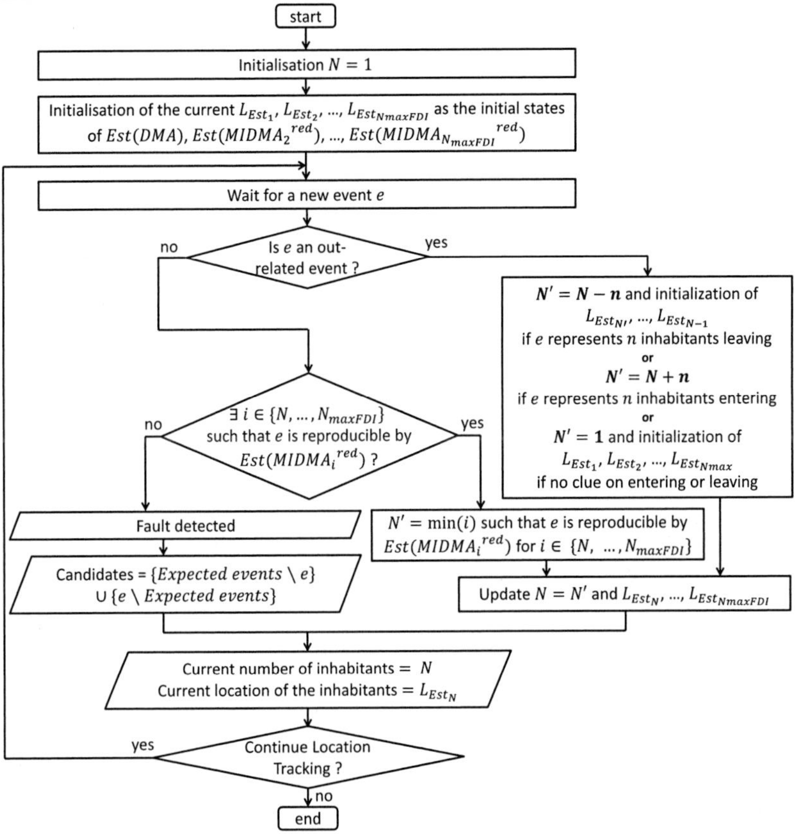

Consequently, Algorithm 3.4 can be applied to perform Location Tracking and sensor FDI in parallel.

**Sensor configuration 1**  The details of the real location, observed event, estimated location and result of the FDI along the scenario for the sensor configuration 1 are given in Table 3.5. Steps 0 and 1 are similar to the fault-free scenario. Then, the unexpected rising edge of the sensor $MD_1$ occurs at step 2. This event is not reproducible in the $DMA$ since there is no transition having state $\mathbf{Z}_7$ as source state and labeled with the event $MD_1\_1$. Thus a fault is detected. Moreover, fault isolation can be performed by computing the residuals. Observed behavior is $MD_1\_1$ and expected behavior is the union of all the events labeling the transitions having $\mathbf{Z}_7$ as source state i.e. expected behavior is $\{MD5\_1, DB_1\_0, DB_1\_1, DB_2\_0, DB_2\_1, MD_2\_1\}$. Thus, the candidates are: $\{MD_1\_1, MD5\_1, DB_1\_0, DB_1\_1, DB_2\_0, DB_2\_1, MD_2\_1\}$. The faulty sensor(s) should be part of the set $Cand = \{MD_1, MD5, DB_1, DB_2, MD_2\}$. In this case, the faulty sensor is $MD_1$ and $MD_1 \in Cand$. However, even if a fault has been detected and isolated, the algorithm continues to run, the estimated location is not updated and remains accurate. The next event to be observed (step 3) is the falling edge of the faulty sensor. Since falling edges are not considered in the model, neither the location is updated, nor a new fault is detected. Then, the following of the scenario remains the same as with the fault-free scenario.

Table 3.5.: Real scenario of motion of a single inhabitant with the first configuration of sensor functioning, Location Tracking and sensor FDI

| Step | Real location $L_{Real}$ | Observed event $e$ | Estimated location $L_{Est}$ | Fault Detection | Fault Isolation |
|---|---|---|---|---|---|
| 0 | $\mathbf{Z}_8$ | $\emptyset$ | $(\mathbf{Z}_1, \mathbf{Z}_2, \mathbf{Z}_3, \mathbf{Z}_4,$ $\mathbf{Z}_5, \mathbf{Z}_6, \mathbf{Z}_7, \mathbf{Z}_8)$ $(ambiguous)$ | no | $\emptyset$ |
| 1 | $\mathbf{Z}_7$ | $MD_5\_1$ | $\mathbf{Z}_7$ $(accurate)$ | no | $\emptyset$ |
| 2 | $\mathbf{Z}_7$ | $MD_1\_1$ | $\mathbf{Z}_7$ $(accurate)$ | yes | $\{MD_1\_1,$ $MD_5\_1,$ $DB_1\_0,$ $DB_1\_1,$ $DB_2\_0,$ $DB_2\_1,$ $MD_2\_1\}$ |
| 3 | $\mathbf{Z}_7$ | $MD_1\_0$ | $\mathbf{Z}_7$ $(accurate)$ | no | $\emptyset$ |
| 4 | $\mathbf{Z}_2$ | $MD_2\_1$ | $\mathbf{Z}_2$ $(accurate)$ | no | $\emptyset$ |
| 5 | $\mathbf{Z}_2$ | $MD_5\_0$ | $\mathbf{Z}_2$ $(accurate)$ | no | $\emptyset$ |
| 6 | $\mathbf{Z}_5$ | $\emptyset$ | $\mathbf{Z}_2$ $(incorrect)$ | no | $\emptyset$ |
| 7 | $\mathbf{Z}_5$ | $MD_2\_0$ | $\mathbf{Z}_2$ $(incorrect)$ | no | $\emptyset$ |
| 8 | $\mathbf{Z}_2$ | $MD_2\_1$ | $\mathbf{Z}_2$ $(accurate)$ | no | $\emptyset$ |
| 9 | $\mathbf{Z}_2$ | $DB_2\_1$ | $(\mathbf{Z}_1, \mathbf{Z}_2)$ $(ambiguous)$ | no | $\emptyset$ |
| 10 | $\mathbf{Z}_1$ | $DB_2\_0$ | $(\mathbf{Z}_1, \mathbf{Z}_2)$ $(ambiguous)$ | no | $\emptyset$ |
| 11 | $\mathbf{Z}_1$ | $MD_1\_1$ | $\mathbf{Z}_1$ $(accurate)$ | no | $\emptyset$ |

**Sensor configuration 2** The details of the real location, observed event, estimated location and result of the FDI along the scenario for the sensor configuration 2 are given in Table 3.6. From step 0 to step 6, the results are exactly the same as in the fault-free case. At step 7, when the inhabitant starts crossing the door between the corridor and the first bedroom, no sensor event is observed because $DB_2$ is stuck at 0. Thus the location is not updated. However and in a non-obvious manner, the current estimated location is accurate whereas it would be ambiguous if the sensor fired (as in the fault-free scenario). In this case, a sensor fault leads to a better Location Tracking result. Of course, this is not always the case, for instance in the next step (step 8), the inhabitant finishes to enter the first bedroom. Once again no sensor event is observed because of the fault of $DB_2$. Thus, the estimated location is not updated and is now incorrect. In this case, the sensor fault leads to a degraded result since the estimated location was just ambiguous in the fault-free scenario. Finally, a rising edge of

$MD_1$ is observed and the estimated location is accurate again. Note that along this scenario and even though a fault occurred in reality, no fault was detected using the Residual-based approach.

Table 3.6.: Real scenario of motion of a single inhabitant with the second configuration of sensor functioning, Location Tracking and sensor FDI

| Step | Real location $L_{Real}$ | Observed event $e$ | Estimated location $L_{Est}$ | Fault Detection | Fault Isolation |
|------|------|------|------|------|------|
| 0 | $\mathbf{Z}_8$ | $\emptyset$ | $(\mathbf{Z}_1, \mathbf{Z}_2, \mathbf{Z}_3, \mathbf{Z}_4,$ $\mathbf{Z}_5, \mathbf{Z}_6, \mathbf{Z}_7, \mathbf{Z}_8)$ $(ambiguous)$ | no | $\emptyset$ |
| 1 | $\mathbf{Z}_7$ | $MD_5\_1$ | $\mathbf{Z}_7$ $(accurate)$ | no | $\emptyset$ |
| 2 | $\mathbf{Z}_2$ | $MD_2\_1$ | $\mathbf{Z}_2$ $(accurate)$ | no | $\emptyset$ |
| 3 | $\mathbf{Z}_2$ | $MD_5\_0$ | $\mathbf{Z}_2$ $(accurate)$ | no | $\emptyset$ |
| 4 | $\mathbf{Z}_5$ | $\emptyset$ | $\mathbf{Z}_2$ $(incorrect)$ | no | $\emptyset$ |
| 5 | $\mathbf{Z}_5$ | $MD_2\_0$ | $\mathbf{Z}_2$ $(incorrect)$ | no | $\emptyset$ |
| 6 | $\mathbf{Z}_2$ | $MD_2\_1$ | $\mathbf{Z}_2$ $(accurate)$ | no | $\emptyset$ |
| 7 | $\mathbf{Z}_2$ | $\emptyset$ | $\mathbf{Z}_2$ $(accurate)$ | no | $\emptyset$ |
| 8 | $\mathbf{Z}_1$ | $\emptyset$ | $\mathbf{Z}_2$ $(incorrect)$ | no | $\emptyset$ |
| 9 | $\mathbf{Z}_1$ | $MD_1\_1$ | $\mathbf{Z}_1$ $(accurate)$ | no | $\emptyset$ |

**Sensor configuration 3** The details of the real location, observed event, estimated location and result of the FDI along the scenario for the sensor configuration 3 are given in Table 3.7. Steps 0 and 1 are the same as in the fault-free scenario. Then, since the sensor $MD_2$ is stuck at 0, no new sensor event is observed when the inhabitant goes in the corridor. Thus, the estimated location is not updated and is now incorrect. No new event is observed while the inhabitant moves in the corridor or in the shower, the estimated location remains incorrect. However, at step 7, when the inhabitant starts to enter the bedroom, he is detected by the door barrier sensor. The estimated location is now updated and is accurate (which is a better result than in the fault-free case). Then, steps 8 and 9 are the same as in the fault-free scenario. Like in the previous sensor configuration, the sensor fault is not detected along the whole scenario. In this case, the impact on the result of Location Tracking is mostly bad, except in step 7.

The conclusion of these three illustrative examples is that the result of Location Tracking is obviously impacted by the potential faults of the sensor. Consequently, the potential occurrence of faults should be taken into account while implementing a Location Tracking algorithm in a real environment. Using the proposed adaption of the residual-based FDI approach is a way to handle potentially faulty sensors but the assumption of a maximal number of inhabitants $N_{maxFDI}$ is hard to guarantee. A new FDI approach devoted to Smart Home should probably be developed because the implications and the issues that are to be faced are not the same as for industrial systems. Still, this study shows the importance of dealing with the potential sensor faults for real implementation of Smart Homes and consequently a discussion on fault-tolerant Location Tracking is proposed in the next section.

Table 3.7.: Real scenario of motion of a single inhabitant with the third configuration of sensor functioning, Location Tracking and sensor FDI

| Step | Real location $L_{Real}$ | Observed event $e$ | Estimated location $L_{Est}$ | Fault Detection | Fault Isolation |
|------|------|------|------|------|------|
| 0 | $\mathbf{Z}_8$ | $\emptyset$ | $(\mathbf{Z}_1, \mathbf{Z}_2, \mathbf{Z}_3, \mathbf{Z}_4,$ $\mathbf{Z}_5, \mathbf{Z}_6, \mathbf{Z}_7, \mathbf{Z}_8)$ (*ambiguous*) | no | $\emptyset$ |
| 1 | $\mathbf{Z}_7$ | $MD_{5}\_1$ | $\mathbf{Z}_7$ (*accurate*) | no | $\emptyset$ |
| 2 | $\mathbf{Z}_2$ | $\emptyset$ | $\mathbf{Z}_7$ (*incorrect*) | no | $\emptyset$ |
| 3 | $\mathbf{Z}_2$ | $MD_{5}\_0$ | $\mathbf{Z}_7$ (*incorrect*) | no | $\emptyset$ |
| 4 | $\mathbf{Z}_5$ | $\emptyset$ | $\mathbf{Z}_7$ (*incorrect*) | no | $\emptyset$ |
| 5 | $\mathbf{Z}_5$ | $\emptyset$ | $\mathbf{Z}_7$ (*incorrect*) | no | $\emptyset$ |
| 6 | $\mathbf{Z}_2$ | $\emptyset$ | $\mathbf{Z}_7$ (*incorrect*) | no | $\emptyset$ |
| 7 | $\mathbf{Z}_2$ | $DB_{2}\_1$ | $\mathbf{Z}_2$ (*accurate*) | no | $\emptyset$ |
| 8 | $\mathbf{Z}_1$ | $DB_{2}\_0$ | $(\mathbf{Z}_1, \mathbf{Z}_2)$ (*ambiguous*) | no | $\emptyset$ |
| 9 | $\mathbf{Z}_1$ | $MD_{1}\_1$ | $\mathbf{Z}_1$ (*accurate*) | no | $\emptyset$ |

### 3.4.4. Discussion on fault-tolerant Location Tracking

The aim of fault-tolerant Location Tracking is to be able to obtain a correct result of Location Tracking despite the potential occurrence of sensor faults. A result is considered to be correct if it is the same (or possibly better) as the result of Location Tracking in the case where no sensor fault occurred.

A solution for fault tolerant Location Tracking consists in performing in parallel the Location Tracking and FDI and after a detected and isolated fault, to stop considering the signals sent by the sensor(s) isolated as faulty. To do it, the FDI algorithm should perform way better as the currently developed Residual-based approach proposed in the previous section because this approach gives too many candidates (thus leading to too many signals not being considered) and is strongly relying on the assumption of a known maximal number of inhabitants (which is hard to guarantee). In addition, as shown for the configurations 2 and 3, there is no guarantee of fault detection using the Residual-based approach.

However, in order to perform fault-tolerant Location Tracking well, a good solution is to install redundant sensors in the Smart Home. Thus, if a fault occurs on one sensor, leading or not to the transmission of an erroneous information, there should be one or several other sensors still giving the correct information.

Fault-tolerant Location Tracking would be performed in the following way. An observed event being non reproducible in $MIDMA_{N_{maxFDI}}^{red}$ means the occurrence of a fault. However, the occurrence of this fault is tolerated. This sensor event is not believed (because not being part of the reproducible behavior) and does not lead to an updated estimation of the location. It is assumed that considering a big number of sensors, if one is faulty, there will always be another one giving correct information and making the model evolve correctly.

However, as said before and to conclude this discussion on fault-tolerant Location Tracking, the aim of this thesis is not to propose efficient FDI approaches for Smart Homes and a fault-tolerant Location Tracking approach. It has just been tried to apply existing approaches from the field of the DES with slight adaption. A whole approach should probably be developed from scratch in order to propose an efficient approach for sensor FDI and fault-tolerant Location Tracking. For this thesis, the contribution is limited to Location Tracking in case of fault-free sensors.

## Conclusion

In this chapter, the procedure to perform online Location Tracking of a known and fixed number of inhabitants, either based on a state estimator built offline or based on an online estimation, has been developed. Moreover, an algorithm for online Location Tracking of an *a priori* unknown number of inhabitants based on the algorithm for known and fixed number of inhabitants has been defined. Finally, considerations relative to the potential faults of the sensors, their detection and isolation as well as fault tolerant Location Tracking have been given.

These algorithms are aimed to give a more or less accurate estimation of the location of the inhabitants. Consequently, the accuracy of model-based Location Tracking has to be evaluated by comparing this estimation with the real location at each time and a guarantee of accurate estimation of the location based on a given model may be proved. These notions are developed in the next chapter.

# 4. Performance evaluation for Location Tracking

## Introduction

In the previous chapters, models of the detectable motion of $N$ inhabitants ($N \in \mathbb{N}^*$) as well as model-based Location Tracking algorithms have been proposed. The illustration of these models and algorithms on examples shows that the result of Location Tracking highly relies on the partition of the house in zones and on the selected instrumentation of the Smart Home. This combination zone partition - instrumentation is more or less explicit, depending on the expert modeling the house. In this chapter, a method to evaluate the performance of the combination zone partition - instrumentation is proposed. An overview of this whole performance evaluation approach is given in Fig. 4.1.

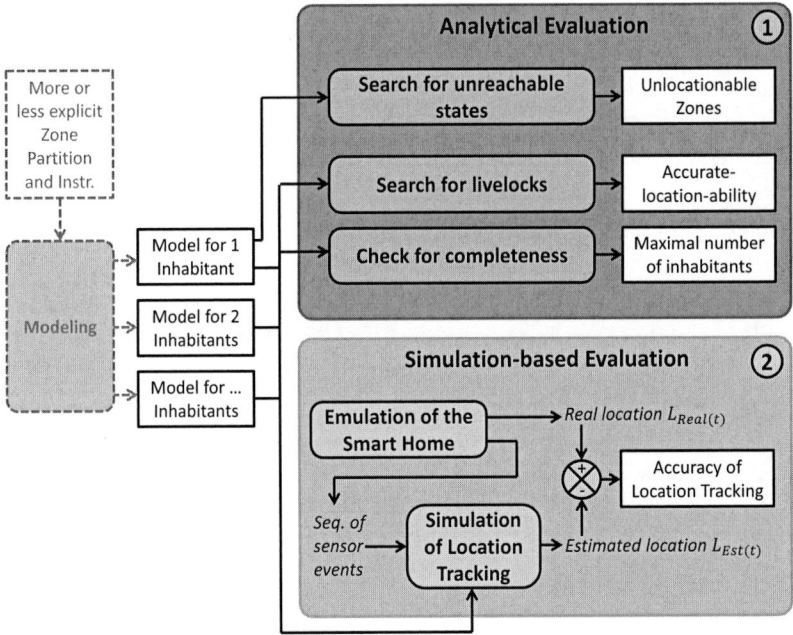

Figure 4.1.: Overview of the whole performance evaluation approach

Based on the models of the detectable motion (model for 1 inhabitant $DMA$, model for 2 inhabitants $MIDMA_2^{red}$, ..., model for $N$ inhabitants $MIDMA_N^{red}$), the evaluation is performed in two steps. The first step consists in the analytical evaluation of the performances. This analytical evaluation is only based on the models (the behavior of the inhabitants is not considered at all) and provides the designer with guarantees of location-ability in the different defined zones of the house or of the maximal number of location-trackable inhabitants. These static criteria can then be used twofold, firstly to guarantee the performance of the combination zone partition - instrumentation, secondly to help the designer to modify the zone partition and/or instrumentation if the results are not satisfying.

In a second step, an evaluation still based on the different models but considering in addition the possible behaviors of the inhabitants is proposed. It provides the designer with dynamic criteria relative to the accuracy of the Location Tracking result for selected scenarios.

The first section of this chapter is dedicated to the analytical performance evaluation and the proposed criteria. In the second section, an iterative procedure to build the model (based on the algorithms of Chapter 2) while performing analytical evaluation is explained. A simulation-based approach for performance evaluation taking the inhabitants' behavior into account is described in the third section. Finally, a procedure to design and improve Smart Homes with the help of the different evaluation criteria is proposed.

## 4.1. Analytical evaluation

This performance evaluation is performed *a priori* i.e. offline and before performing Location Tracking (Danancher et al., 2013c). Thus, different combinations can be evaluated and compared before choosing the best combination zone partition - instrumentation and installing the sensors.

For a given combination zone partition - instrumentation, two criteria are proposed:

- The *unlocationable zones* (inability of the sensors to detect the presence of the inhabitants in certain zones).

- The *strong* and *weak accurate-location-ability*, based on the accuracy of location estimation (guarantee to estimate accurately the location of the inhabitants).

Before detailing these two criteria, the notion of combination zone partition - instrumentation has to be defined. A combination zone partition - instrumentation is denoted as $(P, I)$ where $P$ is the zone partition (formalized with $\mathbf{Z}$ and $Top$ for the considered approach of systematic generation) and $I$ is the instrumentation (formalized with $\mathbf{S}$ and $Obs^+$ for the considered approach of systematic generation). In case of several combinations to be compared, the $l^{th}$ zone partition is denoted as $P^l = (\mathbf{Z}^l, Top^l)$ whereas the $m^{th}$ instrumentation relative to the $l^{th}$ zone partition is denoted as $I^{(m,l)} = (\mathbf{S}^m, Obs^{+(m,l)})$. The combination is denoted as $(P^l, I^{(m,l)})$ and all the models or criteria related to a given combination are annotated with the according exponent $(m, l)$ (for instance $DMA^{(m,l)}$).

### 4.1.1. Unlocationable zones

The ability of a combination zone partition - instrumentation $(P, I)$ to detect the presence of the inhabitant in each zone can be quantified by computing the set of states representing unlocationable zones $Q_{UZ}$ of the automaton $DMA$. They are the states of the $DMA$ being reachable only by the fact that they are initial. It is impossible to detect the presence of the inhabitants in these zones because they do not belong to any state of $Est(DMA)$ or $Est(MIDMA_N^{red}) \ \forall N \geq 2$ except their initial one. Since $MIDMA_N^{red}$ is obtained starting from the $DMA$, the computation of the unlocationable zones is performed only on the $DMA$. The following definition and proposition are proposed for unlocationable zones:

**Definition 16** (Unlocationable Zones). *Unlocationable zones* are particular zones of the home (elements of **Z**) where the inhabitants are never estimated to be in. Each time the inhabitant is really in this zone, his estimated location is incorrect.

**Proposition 3.** The number of unlocationable zones (if they exist) can be quantified by the cardinal of $Q_{UZ}$ with:

$Q_{UZ} = \{q' \in Q \mid \nexists(q, \sigma) \in Q \times \Sigma, \delta(q, \sigma) = q'\}$ with $Q$ the states of the $DMA$.

Since $Q_{UZ} \subseteq Q$, each state $q'$ of $Q_{UZ}$ is related to a zone of **Z** according to Definition 6. Thus $Q_{UZ}$ represents the unlocationable zones.

*Proof.* According to the proposition, a state $q'$ of $Q_{UZ}$ is a state of the $DMA$ being reachable only by the fact that it is initial. Thus, its related zone is related to the initial state of $Est(DMA)$ but it is not related to any other state of $Est(DMA)$ because there is no transition leading to $q'$ in the $DMA$. Consequently, the estimated location will never contain this zone and the inhabitants will never be estimated to be in this zone (except during the initial location estimation).

Moreover, a zone being unlocationable for single inhabitant is also unlocationable for multiple inhabitants. Consequently the set $Q_{UZ}$ is computed using only the $DMA$. □

For the case study of Fig. 2.3 and zone partition of Fig. 2.4, $Q_{UZ} = \{\mathbf{Z}_8; \mathbf{Z}_5\}$ and $|Q_{UZ}| = 2$. This result shows that the inhabitants will never be located in zone $\mathbf{Z}_8$ or in zone $\mathbf{Z}_5$.

This is critical for the zone $\mathbf{Z}_8$ since the location of an inhabitant in this zone is meaning that this inhabitant is not at home. This particular information is mandatory for some applications, for instance applications aiming to detect health problem when the inhabitant is at home. These applications should be deactivated each time the inhabitant is outside in order to avoid false alarms. Thus, the ability to give the information "inhabitant is out of the home" is required. Consequently, this combination $(P, I)$ does not have the ability to guarantee that the apartment is empty.

This can also be considered as critical for the zone $\mathbf{Z}_2$ since it represents the shower. For health problem detection based on local inactivity monitoring, this zone is critical since the inhabitant may slip on the wet floor and fall. Consequently, the ability to detect the inhabitant being in this zone is required since the threshold of acceptable inactivity in these zones is smaller as in other less critical zones (Floeck, 2010).

### 4.1.2. Accurate-location-ability

The accuracy of the location estimation is computed on the state estimator $Est(MIDMA_N^{red})$ ($\forall N \in \mathbb{N}^*$). Note that $MIDMA_1^{red} = DMA$. Each state $q_{Est_N}$ of $Est(MIDMA_N^{red})$ is a subset of the set of states of $MIDMA_N^{red}$. Thus, the accuracy of each estimated location can be quantified by the cardinal of this subset, denoted as $|q_{Est_N}| \; \forall q_{Est_N} \in Q_{Est_N}$, the set of states of $Est(MIDMA_N^{red})$. This notion of estimation accuracy is illustrated on the $DMA$ i.e. for single inhabitant but is similar for $N \geq 2$ inhabitants. For instance in $Est(MIDMA_1^{red}) = Est(DMA)$ of the case study (see $DMA$ and $Est(DMA)$ in Fig. 3.2), $|(\mathbf{Z}_1, \mathbf{Z}_2)| = 2$ and $|\mathbf{Z}_7| = 1$. Each $q_{Est_N} \in Q_{Est_N}$ such that $|q_{Est_N}| = 1$ represents an accurate estimation of the location (inhabitants accurately located each in one zone). In opposite, each $q_{Est_N} \in Q_{Est_N}$ such that $|q_{Est_N}| > 1$ represents a more or less inaccurate estimation of the real location (it cannot be decided accurately in which zone the inhabitants are really located). Based on this consideration, the states of $Est(MIDMA_N^{red})$ ($\forall N \in \mathbb{N}^*$) are divided in two subsets: $Q_{a_N}$

representing the set of accurate estimated locations and $Q_{i_N}$ representing the set of inaccurate estimated locations. $Q_{Est_N} = Q_{a_N} \bigcup Q_{i_N}$ where:

- $Q_{a_N} = \{q_{Est_N} \in Q_{Est_N} \text{ such that } |q_{Est_N}| = 1\}$

- $Q_{i_N} = \{q_{Est_N} \in Q_{Est_N} \text{ such that } |q_{Est_N}| > 1\}$

Note that $Q_{a_N} \cap Q_{i_N} = \emptyset$ and that the initial state $q_{Est_{0_N}} \in Q_{i_N}$. For the case study (see $Est(DMA)$ in Fig. 3.2 (b)), $Q_{a_1} = \{\mathbf{Z}_1, \mathbf{Z}_2, \mathbf{Z}_3, \mathbf{Z}_4, \mathbf{Z}_7\}$ and $Q_{i_1} = \{(\mathbf{Z}_1, \mathbf{Z}_2), (\mathbf{Z}_2, \mathbf{Z}_6), (\mathbf{Z}_1, \mathbf{Z}_2, \mathbf{Z}_3, \mathbf{Z}_4, \mathbf{Z}_5, \mathbf{Z}_6, \mathbf{Z}_7, \mathbf{Z}_8)\}$. The same sets can be calculated for $N > 1$ if the models have been constructed.

Based on these definitions of the sets $Q_{a_N}$ and $Q_{i_N}$, two notions of $N$-accurate-location-ability are defined. They are inspired by the notions of strong and weak detectability of DES (Shu and Lin, 2007).

The *N-accurate-location-ability* is defined as the ability to estimate accurately the location of the $N$ inhabitants. Two definitions for *strong* and *weak N-accurate-location-ability* as well as propositions to check strong or weak $N$-accurate-location-ability are proposed.

**Definition 17** (Strong $N$-accurate-location-ability). A combination $(P, I)$ is strongly $N$-accurate-location-able if after a finite sequence of sensor events the location of the $N$ inhabitants is accurate from now, whatever the inhabitants are doing.

**Proposition 4.** A combination $(P, I)$ is strongly $N$-accurate-location-able if $Q_{a_N}$ is not empty and there is no loop between states of $Q_{i_N}$ and no transition from a state of $Q_{a_N}$ to a state of $Q_{i_N}$. Formally this property is written:

$(P, I)$ is $N$-strongly accurate-location-able if:

$$\begin{cases} \bullet \ Q_{a_N} \neq \emptyset \\ \bullet \ \forall \sigma_1 \sigma_2 \cdots \sigma_m \in \Sigma^* \mid \exists (q_1, q_2, \cdots, q_m, q_{m+1}) \in Q_{i_N}^{m+1}; \\ \quad \delta_{Est_N}(q_1, \sigma_1) = q_2, \cdots, \delta_{Est_N}(q_m, \sigma_m) = q_{m+1}; \\ \quad \forall (q_i, q_j) \in (q_1, q_2, ..., q_{m+1})^2 \ q_i \neq q_j \\ \bullet \ \nexists (q, \sigma) \in Q_{a_N} \times \Sigma \mid \delta_{Est}(q, \sigma) \in Q_{i_N} \end{cases}$$

*Proof.* The condition of non-emptiness of $Q_{a_N}$ and the condition of no transition from $Q_{a_N}$ to $Q_{i_N}$ guarantee that once a state of $Q_{a_N}$ is reached, the subsequent states are also in $Q_{a_N}$ and thus the location will be accurate. Moreover, if a state of $Q_{i_N}$ is reached, the condition of no loop in $Q_{i_N}$ guarantees that, after a sequence of events of maximum length $|Q_{i_N}|$, a state belonging to $Q_{a_N}$ will be reached. $\square$

**Definition 18** (Weak $N$-accurate-location-ability). A combination $(P, I)$ is weakly $N$-accurate-location-able if it is possible that the location of the $N$ inhabitants is accurate from now, depending on what the inhabitants are doing.

**Proposition 5.** A combination $(P, I)$ is weakly $N$-accurate-location-able if $Q_{a_N}$ is not empty and there are loops in $Q_{a_N}$. Formally this property is written:

$(P, I)$ is weakly $N$-accurate-location-able if:

$$\begin{cases} \bullet \ Q_{a_N} \neq \emptyset \\ \bullet \ \exists \sigma_1 \sigma_2 \cdots \sigma_m \in \Sigma^* \mid \exists (q_1, q_2, \cdots, q_m, q_{m+1}) \in Q_{a_N}^{m+1}; \\ \quad \delta_{Est_N}(q_1, \sigma_1) = q_2, \cdots, \delta_{Est_N}(q_m, \sigma_m) = q_{m+1}; \\ \quad \exists (q_i, q_j) \in (q_1, q_2, ..., q_{m+1})^2 \ q_i = q_j \end{cases}$$

*Proof.* The condition of non-emptiness of $Q_{a_N}$ and the fact that each state of $Q_{a_N}$ is accessible (by construction of the estimator) guarantees that it is possible to enter at least one loop on $Q_{a_N}$. Thus, a sequence of events exists such that, after a certain number of events, the current location and subsequent locations become accurate. □

For the case study, strong 1-location-ability is not obtained because there is at least one transition between $Q_{a_1}$ and $Q_{i_1}$ in $Est(DMA)$ (for example the transition from $\mathbf{Z}_1$ to $(\mathbf{Z}_1, \mathbf{Z}_2)$ labeled with event $DB_2\_1$). Thus a sequence of events exists such that the estimated location becomes inaccurate after it has been accurate. For instance the sequence of events $MD_1\_1, DB_2\_1$ leads to the sequence of estimated locations $(\mathbf{Z}_1, \mathbf{Z}_2, \mathbf{Z}_3, \mathbf{Z}_4, \mathbf{Z}_5, \mathbf{Z}_6, \mathbf{Z}_7, \mathbf{Z}_8)$, $\mathbf{Z}_1, (\mathbf{Z}_1, \mathbf{Z}_2)$ where $\mathbf{Z}_1$ is accurate and the first and third estimated locations are inaccurate.

However, since $Q_{a_N}$ is not empty and there exist loops on $Q_{a_N}$ (for instance the self-loop on state $\mathbf{Z}_1$, formally written $\delta_{Est_1}(\mathbf{Z}_1, MD_1\_1) = \mathbf{Z}_1$), the weak 1-location-ability is guaranteed. At least one sequence of events exists such that the estimated location becomes accurate and all the subsequent estimated locations are accurate. For instance the sequence $DB_2\_1, MD_1\_1, MD_1\_1^\star$ leads the estimated location to be inaccurate in $(\mathbf{Z}_1, \mathbf{Z}_2, \mathbf{Z}_3, \mathbf{Z}_4, \mathbf{Z}_5,$ $\mathbf{Z}_6, \mathbf{Z}_7, \mathbf{Z}_8)$, then inaccurate in $(\mathbf{Z}_1, \mathbf{Z}_2)$, finally accurate in $\mathbf{Z}_1$ and it remains accurate in $\mathbf{Z}_1$ since $\delta_{Est_1}(\mathbf{Z}_1, MD_1\_1) = \mathbf{Z}_1$.

### 4.1.3. Maximal number of trackable inhabitants

Considering the proposed algorithm for Location Tracking of an *a priori* unknown number of inhabitants (Algorithm 3.3), a maximal number of trackable inhabitants can be defined. This number is denoted as $N_{max}$. According to Algorithm 3.3, the current number of inhabitants $N$ is determined thanks to the events being non-reproducible in a model $Est(MIDMA_{N-1}^{red})$. Consequently, if all the events are reproducible from all the states of a certain model of $N$ inhabitants, it will never be possible that the current number of inhabitants becomes $N + 1$. This leads to the following definitions and proposition:

**Definition 19** (Maximal number of trackable inhabitants $N_{max}$). The maximal number of trackable inhabitants is the maximal estimated number of inhabitants that can be given while performing Location Tracking of an *a priori* unknown number of inhabitants.

**Definition 20** (Complete Finite Automaton). A Finite Automaton $Aut = (Q, \Sigma, \delta, Q_0)$ is said to be complete if, from each state, a transition labeled with each of the events exists. Formally:

$Aut$ is a complete Finite Automaton if $\forall(q, \sigma) \in Q \times \Sigma \ \delta(q, \sigma)!$

**Proposition 6.** The maximal number of trackable inhabitants $N_{max}$ is defined by the first complete finite automaton representing the detectable motion of several inhabitants. Formally:

- if $Est(DMA)$ is complete, $N_{max} = 1$

- else $N_{max} = \min_{i>1}(i)$ such that $Est(MIDMA_i^{red})$ is complete and $Est(MIDMA_{i-1}^{red})$ is not complete.

*Proof.* Based on Definition 20, if $i$ is such that $Est(MIDMA_i^{red})$ is a complete finite automaton, it is impossible to observe a behavior not being reproducible by $Est(MIDMA_i^{red})$. Consequently, based on the algorithm for Location Tracking of an *a priori* unknown number of inhabitants (Algorithm 3.3), it is impossible to increase the number of inhabitants by observing a non reproducible behavior. Thus, $N_{max}$ is the minimal value of $i$ for which $Est(MIDMA_i^{red})$ is complete. □

For the reduced case study (first bedroom, corridor, second bedroom and bathroom of Fig. 2.3), $Est(DMA)$ is not a complete automaton and $Est(MIDMA_2^{red})$ (see Fig. 3.3) is a complete automaton. Thus the maximal number of trackable inhabitants $N_{max}$ is equal to 2. Consequently, if the number of inhabitants in the house becomes strictly greater than 2, the result of the Location Tracking will potentially be false. Furthermore, there is no need to compute the models for $N > 2$ because according to Algorithm 3.3 they will never be used and thus are useless. For the whole home (case study of Fig. 2.3), $N_{max} = 4$.

Based on this whole analytical approach, a procedure to build and evaluate the models in parallel is proposed in the next section.

## 4.2. Iterative procedure for model-building and analytical evaluation

Since $N_{max}$ is not known *a priori*, Algorithm 4.1 is proposed. It is aimed to build all the models required for the Location Tracking of multiple inhabitants without building useless models ($MIDMA_N^{red}$ for $N > N_{max}$).

**Algorithm 4.1** Algorithm for iterative model-building and analytical evaluation

In a first step, the $DMA$ is built and analytical evaluation is performed for Single Inhabitant Location Tracking ($i = 1$). Moreover, based on the potential completeness of the model, it is checked if $N_{max} = 1$ or not. If not, then $i$ is increased, the model for $i$ inhabitants $MIDMA_i^{red}$ is built and the evaluation for $i$-Inhabitants Location Tracking is performed. In addition, based on the potential completeness of the model, it is checked if $N_{max} = i$ or not. If not, then $i$ is increased and model building and evaluation continue. If $N_{max} = i$, then $N_{max}$ is found and the algorithm stops. At the end, all the required models (and no more) are built and the results of the analytical evaluation (*unlocationable zones, Acurate-Location-Ability, maximal number of trackable inhabitants*) are all already obtained.

Once the models have been built and used for analytical evaluation of the performances, a second phase of evaluation based on the simulation can be performed. This simulation-based evaluation is presented in the next section.

## 4.3. Simulation-based evaluation

The previously proposed analytical approach allows determining asymptotic performance criteria such as the set of unlocationable zones (if they exist) and the *accurate-location-ability*. Nevertheless, this static performance evaluation does not take into account the human behavior, which is obviously complex, since neither deterministic nor stochastic but rather arbitrary and potentially irrational. Thus, in this section, it is proposed to consider the human behavior and to define performance criteria depending on this behavior (Danancher et al., 2013a).

There are several possibilities to take the human behavior into account in order to evaluate the performance of a Location Tracking algorithm and of the combination zone partition - instrumentation. An overview of the proposed approach is given in Fig. 4.2. Obviously, the most favorable case consists in building the Smart Home with the chosen instrumentation and in observing real people living inside the house. Thus a real human behavior (with all its uncertainty, non-determinism and arbitrariness) is considered. However, in order to evaluate the Location Tracking procedure, both the real location and the estimated location of the inhabitants are assumed to be known at each time. Even if the estimated location is always known because it is given by the Location Tracking algorithm based on the observed sensor events, the real location is quite difficult to monitor online. Another reliable Location Tracking method should run in parallel (involving wearable sensors or video camera) for instance. The intrusive aspects of these other methods as well as their costs make them not interesting for a quick evaluation of a combination zone partition - instrumentation.

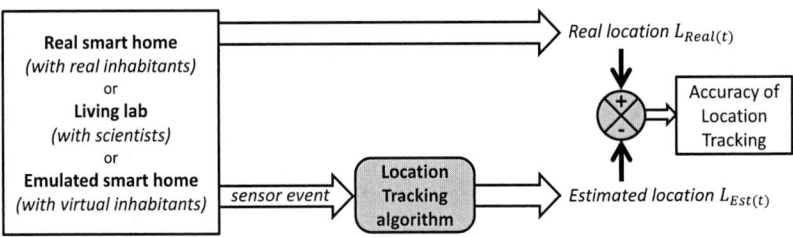

Figure 4.2.: Overview of the possible human-behavior-based performance evaluation

Another possibility consists in building a living lab, similar to the real home but with scientists living inside. In this case, the intrusiveness is no longer a real problem and video camera can be set up everywhere in order to get the real location of the inhabitants. However, several drawbacks have to be noted. It is still difficult because time consuming to monitor online the real location of the inhabitants. Moreover, if several instrumentations are to be compared, it would be hard or even impossible for the people living there to perform several times exactly the same routine in order to compare different instrumentations.

Finally, the last possibility is to consider an emulated Smart Home. There is in this case no problem of cost or of intrusiveness since the considered Smart Home is a virtual one. Different instrumentations can easily be considered. Still the problem of the realism of the human behavior in this emulated house is a problem, since it is not a real behavior as in a real home with real inhabitants. However, it allows performing quickly (also by compressing the time, by playing the motion of the inhabitants faster than in real time) a human-behavior-based evaluation of a combination zone partition - instrumentation.

An overview of the proposed approach is given in Fig. 4.3. The simulation of Location Tracking is performed using the algorithms of Chapter 3, i.e. the simulator is based on exactly the same algorithm as the one used for Location Tracking in the real home. Moreover, this simulation is aiming at evaluating the relevance of the FA model ($DMA$ or $MIDMA_N^{red}$) as it will be used for performing the Location Tracking. At this point, for a given sequence of sensor events, there is no deviation between the Location Tracking done by simulation and the expected Location Tracking in real homes. The goal of the simulation is to evaluate *a priori* the relevance of the model of the detectable motion for Location Tracking by generating a lot of relevant sequences of events that represent accurately the motions and actions of the inhabitants, without having to perform tests in the real instrumented environment. The emulated behavior is assumed to be close to a real behavior of the inhabitants in the Smart Home.

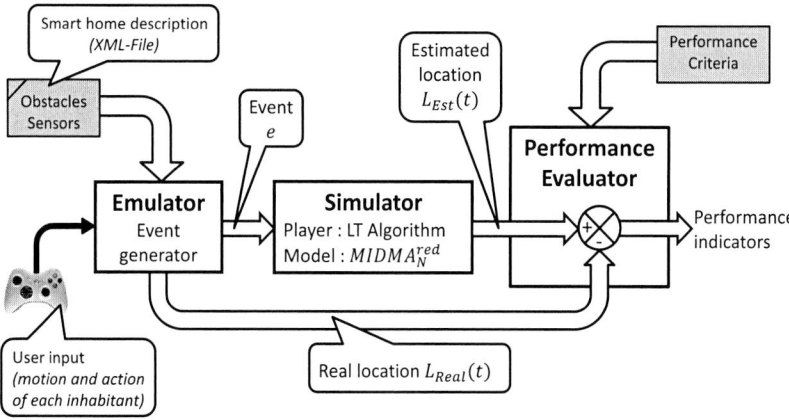

Figure 4.3.: Overview of the procedure for simulation-based performance evaluation

An emulator which allows immersing a user (thanks to a keyboard or a joystick) in a virtual environment reproducing the topology and the instrumentation of the Smart Home has been developed (details relative to the implementation are given in the appendix of this thesis). The advantage of the developed emulator is that the behavior of the inhabitant is not modeled. Approaches to model the human behavior exist (Tan, 2007; Abdul Majid, 2011) but in the proposed approach only the environment (walls, doors, sensors) is modeled. The behavior of the inhabitant is a real human behavior of a real human using the joystick to immerse himself in the emulated Smart Home. The aim of this approach is to recreate the real behavior, the most faithfully as possible, without modeling it.

The sensors are reacting to the inhabitants' motion and action and provide the according sequence of events. Furthermore, in the emulator, the exact location of each inhabitant at each time is known. It is then possible to evaluate the dynamic performance of the accuracy of the Location Tracking model by comparing the estimated location with the real one.

Several criteria are defined in order to evaluate the results of the simulated Location Tracking. The evaluation criteria are based on the confusion matrix (Kubat et al., 1998). A confusion matrix represents the number of times a prediction corresponds or not to a real situation. It is mainly used to evaluate data-driven learning approaches e.g. Activities of Daily Living recognition in Smart Homes (Chikhaoui et al., 2011). Note that these criteria are not exclusively dedicated to simulation-based evaluation, they can also be favorably used when considering a real behavior from a real home or the behavior observed in a living lab (as in Fig. 4.2).

A confusion matrix called $CM_{LT}$ is defined in order to evaluate the accuracy of the estimated location of inhabitants compared to their real location. Considering either the case of single inhabitant, or of $N$ inhabitants, or of *a priori* unknown number of inhabitants, there are $|L|$ possible locations where:

- $|L| = |\mathbf{Z}| = |Q|$, the states of $DMA$ for single inhabitant.

- $|L| = |Q_N|$, the states of $MIDMA_N^{red}$ for $N$ inhabitants.

- $|L| = |Q| + \sum\limits_{i=2}^{N_{max}} |Q_i|$ for an *a priori* unknown number of inhabitants (where the maximal number of inhabitants is $N_{max}$).

Consequently, the expert has to choose the confusion matrix fitting the case he wants to evaluate (Single Inhabitant Location Tracking, $N$-Inhabitants Location Tracking or *a priori* unknown number of inhabitants Location Tracking). In any case, there are two possible definitions of the confusion matrix. The first definition is time-based and is given below. The second definition is event-driven and is given after developing all the criteria derived from the time-based confusion matrix.

### 4.3.1. Time-based confusion matrix and derived criteria

The time-based confusion matrix $CM_{LT}(T)$ is a $|L| \times |L|$-matrix and each $CM_{LT_{(i,j)}}(T)$ is defined as follow:

$$CM_{LT_{(i,j)}}(T) = \frac{1}{T} \int_0^T \frac{\mathbf{1}_{\{(L_i=L_{Real}(t)) \ \wedge \ (L_j \in L_{Est}(t))\}}}{|L_{Est}(t)|} dt \qquad (4.1)$$

where:

- $T$ is the duration of the simulation,

- $\mathbf{1}_{\{predicate\}} = 1$ if *predicate* is true and 0 otherwise,

- $L_i$ (resp. $L_j$) represents the $i^{th}$ (resp. the $j^{th}$) possible location of the inhabitants,

- $L_{Real}(t)$ is the real location (each in one zone) of the inhabitants at the date $t$,

- $L_{Est}(t)$ is the estimated location (in a set of zones, possibly containing only one combination of zones) of the inhabitants at the date $t$,

- $|L_{Est}(t)|$ is the number of combination of zones composing the estimated location (for instance, if the estimated location $L_{Est}(t_1) = (\mathbf{Z}_1, \mathbf{Z}_2)$, then $|L_{Est}(t_1)| = 2$).

Note that the sum of all the $CM_{LT_{(i,j)}}$ is equal to 1 i.e.

$$\forall T \sum_{i,j} CM_{LT_{(i,j)}}(T) = 1 \tag{4.2}$$

Three purely numerical examples of confusion matrices with no link to a real case study are given below. Each matrix represents a possible use of this matrix (single inhabitant $CM_{LT}^{SILT}$, known number of inhabitants $CM_{LT}^{N-ILT}$, a priori unknown number of inhabitants $CM_{LT}^{MILT}$) for a partition in 3 zones.

$CM_{LT}^{SILT}$ should be used to perform evaluation for single inhabitant:

$$CM_{LT}^{SILT}(T_1) = \begin{bmatrix} 0.2 & 0.1 & 0 \\ 0 & 0.3 & 0.1 \\ 0.05 & 0.05 & 0.2 \end{bmatrix} =$$

|  |  | estimated location $L_j$ | | |
|---|---|---|---|---|
|  |  | $\mathbf{Z}_1$ | $\mathbf{Z}_2$ | $\mathbf{Z}_3$ |
| real | $\mathbf{Z}_1$ | 0.2 | 0.1 | 0 |
| location | $\mathbf{Z}_2$ | 0 | 0.3 | 0.1 |
| $L_i$ | $\mathbf{Z}_3$ | 0.05 | 0.05 | 0.2 |

This matrix should be understood in the following way.

- During 30% of the simulation time, the real location was $\mathbf{Z}_2$ and the estimated location was $\mathbf{Z}_2$ or contained $\mathbf{Z}_2$ ($CM_{LT_{(2,2)}}^{SILT}(T_1) = 0.3$).

- During 5% of the simulation time, the real location was $\mathbf{Z}_3$ and the estimated location was $\mathbf{Z}_1$ or contained $\mathbf{Z}_1$, i.e. the inhabitant's location was incorrectly estimated ($CM_{LT_{(3,1)}}^{SILT}(T_1) = 0.05$).

- When the inhabitant is in $\mathbf{Z}_1$, he is never estimated as being in $\mathbf{Z}_3$ ($CM_{LT_{(1,3)}}^{SILT}(T_1) = 0$)

$CM_{LT}^{2-ILT}$ should be used to perform evaluation for 2 inhabitants:

$$CM_{LT}^{2-ILT}(T_1) = \begin{bmatrix} 0.2 & 0.05 & 0.02 & 0.01 & 0 & 0.02 \\ 0.01 & 0.15 & 0.02 & 0.01 & 0 & 0.01 \\ 0 & 0 & 0.1 & 0.02 & 0 & 0.03 \\ 0.03 & 0.02 & 0 & 0.05 & 0 & 0 \\ 0.05 & 0 & 0.05 & 0 & 0 & 0 \\ 0.02 & 0.03 & 0 & 0 & 0 & 0.1 \end{bmatrix} \begin{array}{l} (L_1 = \mathbf{Z}_1.\mathbf{Z}_1) \\ (L_2 = \mathbf{Z}_1.\mathbf{Z}_2) \\ (L_3 = \mathbf{Z}_1.\mathbf{Z}_3) \\ (L_4 = \mathbf{Z}_2.\mathbf{Z}_2) \\ (L_5 = \mathbf{Z}_2.\mathbf{Z}_3) \\ (L_6 = \mathbf{Z}_3.\mathbf{Z}_3) \end{array}$$

Finally $CM_{LT}^{MILT}$ should be used to perform evaluation for an a priori unknown number of inhabitants up to 2:

$$CM_{LT}^{MILT}(T_1) = \begin{bmatrix} 0.02 & 0.01 & 0.01 & 0 & 0 & 0 & 0 & 0 & 0 \\ 0 & 0.04 & 0.02 & 0 & 0 & 0 & 0 & 0 & 0 \\ 0 & 0 & 0.01 & 0 & 0 & 0 & 0 & 0 & 0 \\ 0 & 0.01 & 0 & 0.1 & 0.05 & 0.02 & 0.01 & 0 & 0.02 \\ 0 & 0 & 0 & 0.01 & 0.15 & 0.02 & 0.01 & 0 & 0.01 \\ 0 & 0 & 0.01 & 0 & 0 & 0.1 & 0.02 & 0 & 0.03 \\ 0.02 & 0 & 0 & 0.03 & 0.02 & 0 & 0.05 & 0 & 0 \\ 0 & 0 & 0 & 0.05 & 0 & 0.05 & 0 & 0 & 0 \\ 0.01 & 0 & 0 & 0.02 & 0.03 & 0 & 0 & 0 & 0.04 \end{bmatrix} \begin{array}{l} (L_1 = \mathbf{Z}_1) \\ (L_2 = \mathbf{Z}_2) \\ (L_3 = \mathbf{Z}_3) \\ (L_4 = \mathbf{Z}_1.\mathbf{Z}_1) \\ (L_5 = \mathbf{Z}_1.\mathbf{Z}_2) \\ (L_6 = \mathbf{Z}_1.\mathbf{Z}_3) \\ (L_7 = \mathbf{Z}_2.\mathbf{Z}_2) \\ (L_8 = \mathbf{Z}_2.\mathbf{Z}_3) \\ (L_9 = \mathbf{Z}_3.\mathbf{Z}_3) \end{array}$$

Note that in both matrices $CM_{LT}^{2ILT}(T_1)$ and $CM_{LT}^{MILT}(T_1)$ the inhabitants will never be located in $\mathbf{Z_2.Z_3}$ because the corresponding column (the last but one column) in the matrix is full of 0. Consequently, this location will always be incorrect. Note also that the cases of accurate locations correspond to the diagonal elements of the matrix.

Based on this matrix, several performance criteria can be defined. The first one is named *accuracy* and gives the proportion of time for which the estimated location was the real one during the simulation (e.g. if the estimated location was correct during 7 minutes for a simulation of 10 minutes, then the accuracy is equal to 70%). Formally it is the sum of the diagonal elements of the confusion matrix:

$$accuracy(T) = \sum_i CM_{LT_{(i,i)}}(T) \in [0, 1] \qquad (4.3)$$

The *accuracy* gives a global measure of the Location Tracking performance. For the example of $CM_{LT}^{SILT}(T_1)$, $accuracy(T_1) = 0.2 + 0.3 + 0.2 = 0.7$.

Complementary to this indicator, two criteria concerning each possible location are defined. The *precision* $p$ is the proportion of time for which the estimated location correctly represents the real location and the *recall* $r$ is the proportion of time for which a real location is correctly estimated. Formally, for a given location $i$:

$$p(i, T) = \frac{CM_{LT_{(i,i)}}(T)}{\displaystyle\sum_j CM_{LT_{(j,i)}}(T)} \in [0, 1] \qquad (4.4)$$

$$r(i, T) = \frac{CM_{LT_{(i,i)}}(T)}{\displaystyle\sum_j CM_{LT_{(i,j)}}(T)} \in [0, 1] \qquad (4.5)$$

For the example of $CM_{LT}^{SILT}(T_1)$, the precisions are $p(1, T_1) = \dfrac{0.2}{0.2 + 0 + 0.05} = 0.8$, $p(2, T_1) = 0.67$, $p(3, T_1) = 0.67$ and the recalls are $r(1, T_1) = \dfrac{0.2}{0.2 + 0.1 + 0} = 0.67$, $r(2, T_1) = 0.75$, $r(3, T_1) = 0.67$.

The criteria *precision* and *recall* can be combined using a geometric mean called *gmean*:

$$gmean(i, T) = \sqrt{r(i, T) \cdot p(i, T)} \in [0, 1] \qquad (4.6)$$

Again for the example of $CM_{LT}^{SILT}(T_1)$, $gmean(1, T_1) = \sqrt{r(1, T_1) \cdot p(1, T_1)} = \sqrt{0.8 \times 0.67} = 0.73$, $gmean(2, T_1) = 0.71$ and $gmean(3, T_1) = 0.67$. This geometric mean gives a unique evaluation of the performance for each possible location.

These criteria are all based on a time-based confusion matrix $CM_{LT}(T)$ for which the assumptions that the time of the simulation is directly proportional to the real time holds. However, it is also possible to consider an event-driven confusion matrix as defined below.

### 4.3.2. Event-driven confusion matrix and derived criteria

The event-driven confusion matrix $CM_{LT}(E)$ is a $|L| \times |L|$-matrix and each $CM_{LT_{(i,j)}}(E)$ is defined as follow:

$$CM_{LT_{(i,j)}}(E) = \frac{1}{|E|} \sum_{k=0}^{|E|} \frac{\mathbf{1}_{\{(L_i = L_{Real}(k)) \, \wedge \, (L_j \in L_{Est}(k))\}}}{|L_{Est}(k)|} dt \qquad (4.7)$$

91

where:

- $E$ is the sequence of changes of real or estimated location during the simulation. Note that this sequence is not the same as the sequence of sensor events,

- $\mathbf{1}_{\{predicate\}} = 1$ if $predicate$ is true and 0 otherwise,

- $L_i$ (resp. $L_j$) represents the $i^{th}$ (resp. the $j^{th}$) possible location of the inhabitants,

- $L_{Real}(k)$ is the real location (each in one zone) of the inhabitants after the $k^{th}$ change of location,

- $L_{Est}(k)$ is the estimated location (in a set of zones, possibly containing only one combination of zones) of the inhabitants after the $k^{th}$ change of location,

- $|L_{Est}(k)|$ is the number of combination of zones composing the estimated location (for instance, if the estimated location $L_{Est}(k_1) = (\mathbf{Z}_1, \mathbf{Z}_2)$, then $|L_{Est}(k_1)| = 2$).

Based on this alternative confusion matrix, *accuracy*, *precision*, *recall* and *gmean* are computed in a similar manner.

$$accuracy(E) = \sum_i CM_{LT_{(i,i)}}(E) \in [0,1] \tag{4.8}$$

$$p(i,E) = \frac{CM_{LT_{(i,i)}}(E)}{\sum_j CM_{LT_{(j,i)}}(E)} \in [0,1] \tag{4.9}$$

$$r(i,E) = \frac{CM_{LT_{(i,i)}}(E)}{\sum_j CM_{LT_{(i,j)}}(E)} \in [0,1] \tag{4.10}$$

$$gmean(i,E) = \sqrt{r(i,E) \cdot p(i,E)} \in [0,1] \tag{4.11}$$

Based on these two definitions of the confusion matrix, the expert has to choose between the time-based confusion matrix $CM_{LT}(T)$ or the event-driven confusion matrix $CM_{LT}(E)$. From a DES point of view, the event-driven matrix seems better since it is only based on events (in this case, events representing a change of location, estimated or real) and it does not take into account the timed behavior of the inhabitant at all. However, from a Smart Home point of view, for some applications, the important point is the behavior of the inhabitants and the reaction of the Location Tracking algorithms to this behavior. In this latter case, considering the time is important and the time-based confusion matrix would be well adapted. To conclude on the choice of time-based or event-driven criteria, this choice has to be made according to the expected usage of Location Tracking (smart energy monitoring, health problem detection, etc...). There is no strong conclusion on one being better than another. In the following, only the time-based criteria will be considered even though the other ones can also be implemented.

All of these criteria (Confusion matrix, *accuracy*, *precision*, *recall* and *gmean*) are calculated and displayed online while simulating the behavior of one or several inhabitants in the Smart Home. This helps the designer to visualize the weakness of his chosen combination zone partition - instrumentation and gives him instantaneously a feedback on the scenario he is playing.

### 4.3.3. Illustration on a case study

The results of the application of this simulation-based evaluation approach applied on the case study of Fig. 2.3 and zone partition of Fig. 2.4 are given below. One scenario involving a single inhabitant moving in the house during 5 minutes has been simulated. At the end of the simulation (i.e. after $T_1 = 5$ minutes), the confusion matrix of Single Inhabitant Location Tracking $CM_{LT}^{SILT}(T_1)$ is the following:

$$CM_{LT}^{SILT}(T_1) = \begin{bmatrix} 0.239 & 0.004 & 0.000 & 0.000 & 0.000 & 0.000 & 0.000 & 0.000 \\ 0.003 & 0.074 & 0.001 & 0.000 & 0.000 & 0.015 & 0.008 & 0.000 \\ 0.000 & 0.008 & 0.029 & 0.002 & 0.000 & 0.000 & 0.033 & 0.000 \\ 0.000 & 0.000 & 0.000 & 0.075 & 0.000 & 0.000 & 0.055 & 0.000 \\ 0.000 & 0.029 & 0.000 & 0.000 & 0.000 & 0.029 & 0.000 & 0.000 \\ 0.000 & 0.026 & 0.000 & 0.000 & 0.000 & 0.026 & 0.000 & 0.000 \\ 0.000 & 0.003 & 0.000 & 0.000 & 0.000 & 0.000 & 0.146 & 0.000 \\ 0.013 & 0.012 & 0.012 & 0.012 & 0.012 & 0.012 & 0.123 & 0.000 \end{bmatrix} \begin{matrix} (L_1 = \mathbf{Z}_1) \\ (L_2 = \mathbf{Z}_2) \\ (L_3 = \mathbf{Z}_3) \\ (L_4 = \mathbf{Z}_4) \\ (L_5 = \mathbf{Z}_5) \\ (L_6 = \mathbf{Z}_6) \\ (L_7 = \mathbf{Z}_7) \\ (L_8 = \mathbf{Z}_8) \end{matrix}$$

Moreover, the *accuracy* has been calculated at each time and its evolution is shown in Fig. 4.4 (a). At the end of the scenario, the *accuracy* is almost 60%, a not so high value that confirms the analytical evaluation (poor results of the analytical evaluation since there are 2 unlocationable zones).

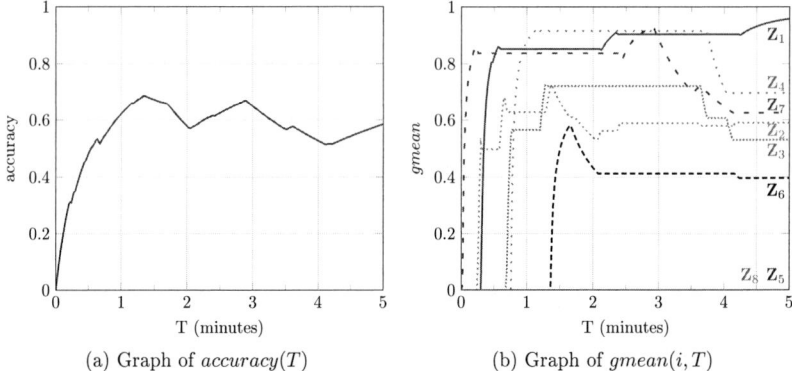

(a) Graph of *accuracy*$(T)$       (b) Graph of *gmean*$(i, T)$

Figure 4.4.: Evolution of the overall *accuracy* and of the local performances using *gmean*$(i, T)$ along the time $T$

The *accuracy* is a global indicator of the performance, to have more detailed results relative to the Location Tracking for each zone, the evolution of the criterion *gmean* is shown in Fig. 4.4 (b) for each zone of the house. Again, the results of the analytical evaluation are confirmed, the *gmean* value for $\mathbf{Z}_5$ and $\mathbf{Z}_8$ is stuck to 0 since they are the unlocationable zones. However, other conclusion can be drawn from these simulation-based results that cannot be obtained only with the analytical evaluation. The best results are obtained for the zone $\mathbf{Z}_1$ that means that the instrumentation in this room is well chosen and placed. But zones $\mathbf{Z}_6$ and $\mathbf{Z}_3$ for instance show low performances, meaning that a sensor is probably not well chosen (probably the door barrier sensor is not the best choice for zone $\mathbf{Z}_6$) or placed (maybe the motion detector in zone $\mathbf{Z}_3$ is too far from the door).

To conclude this illustration of the simulation-based evaluation, note that it provides the designer with results and indicators that are complementary to the results of the analytical evaluation procedure. However, the simulation-based results are strongly depending on the simulated scenario. Thus, for two different scenarios, the results may be very different. This approach should be used to test particularly critical scenarios or to compare several instrumentations on a same scenario (as it is shown in the next section).

## 4.4. Evaluation-aided improvement of Smart Home's instrumentations

### 4.4.1. Overview of the proposed approach

Based on the previously described evaluation procedures (analytical and simulation-based), an approach for assisted improvement of Smart Home instrumentation is proposed. This approach is named evaluation-aided improvement of Smart Home instrumentation and an overview is given in Fig. 4.5.

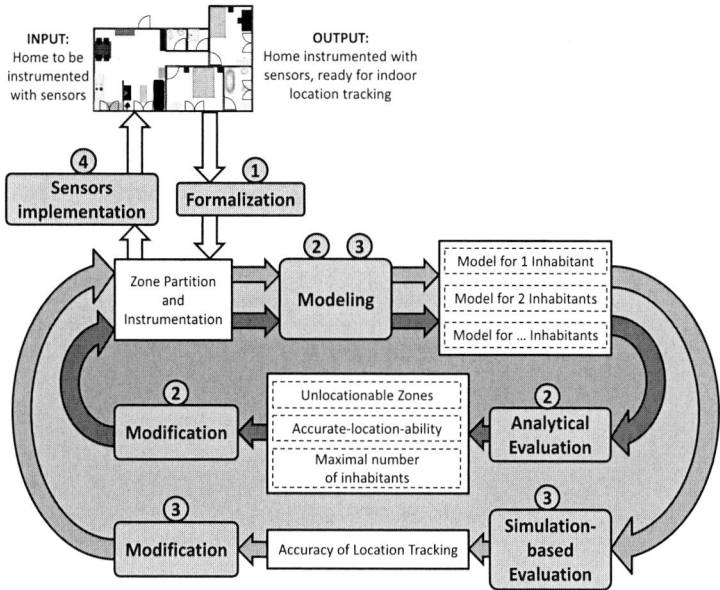

Figure 4.5.: Overview of the approach for evaluation-aided improvement of Smart Home instrumentation

The idea behind is to provide the designer with indications to improve his choice in terms of zone partition and instrumentation in order to finally get a Smart Home and models compliant with his needs. A closed loop of improvement is proposed. First, the designer describes an initial combination zone partition - instrumentation (step 1 of Fig. 4.5), the models are systematically generated and evaluated using the iterative procedure of Algorithm. 4.1 (modeling and analytical evaluation steps of Fig. 4.5). Then, based on these results, the designer can make any changes he wants in the combination zone partition - instrumentation (modification step of Fig. 4.5) and compute again the model building and analytical evaluation. Modeling, analytical evaluation and modification constitute the loop 2 in Fig. 4.5 which is the first loop of improvement.

When the analytical results are satisfying or when several possible combinations having the same analytical performances are to be compared, the designer can perform simulation-based evaluation and test some particular or critical scenarios (loop 3, composed of modeling, simulation-based evaluation and modification). Different combinations can be compared on exactly the same scenario. Thus, either the designer is satisfied with one of the tested combinations and choose this one for real implementation (step 4), or he gets indications to improve again the combination and the loop starts again.

In any case, it is interesting to first try to solve the problem of Single Inhabitant Location Tracking and get a satisfying instrumentation for this case before trying to solve the Multiple Inhabitants Location Tracking problem.

Consequently, the possible actions of the designer to modify the combination are given below in the order they should be performed.

1. Evaluation for Single Inhabitant Location Tracking

   a) Based on the results of the analytical evaluation

      - If there are unlocationable zones (i.e. $Q_{UZ} \neq \emptyset$), then sensor(s) must be added in the according zone(s).

      - If $(\mathbf{Z}_i.\mathbf{Z}_j) \in Q_i \wedge \mathbf{Z}_i \in Q_a \wedge \mathbf{Z}_j \in Q_a$, then there is probably an exceeding sensor (typically a door sensor between the two zones $\mathbf{Z}_i$ and $\mathbf{Z}_j$ already observed each by another sensor).

      - If $(\mathbf{Z}_i.\mathbf{Z}_j) \in Q_i \wedge \mathbf{Z}_i \notin Q_a$, then modify the zone partition by merging the zones $\mathbf{Z}_i$ and $\mathbf{Z}_j$ (this solution should be avoided in order to keep as much as possible the choice of the designer for the zone partition) or modify the placement of existing sensors and add new sensors in zone $\mathbf{Z}_i$.

   b) Based on the results of the simulation-based evaluation

      - If the overall *accuracy* is low, have a closer look on *gmean* for the different zones.

      - If $gmean(i, T)$ is too low, add a sensor in zone $\mathbf{Z}_i$.

2. Evaluation for Multiple Inhabitants Location Tracking

   a) Based on the results of the analytical evaluation

      - If $N_{max}$ is too low, split existing zones into more zones. Adapt the instrumentation to this new zone partition, probably by adding new sensors.

   b) Based on the results of the simulation-based evaluation

      - If the overall *accuracy* is low, have a closer look on *gmean* for the different locations.

      - If $gmean(i, T)$ is too low, add sensors in the zones composing the location $L_i$ .

To summarize this, the evaluation-aided improvement approach has two parameters: the zone partition and the instrumentation, and several evaluation criteria: analytical criteria (unlocationable zones, accurate-location-ability, maximal number of trackable inhabitants) and simulation-based criteria (global and local accuracy). The aim of this procedure is to improve the Smart Home in the sense of the evaluation criteria by acting primarily on the instrumentation. Indeed, the different applications of Location Tracking highly rely on the zone partition; consequently this parameter should be modified only if it does not impact on the quality of the application. As an example, for health problem detection based on Location Tracking and local inactivity monitoring, some zones must not be modified at all (particularly they cannot be merged with other zones because they are critical for some health problems, the bathroom being a good example of these non-modifiable zones. In these cases, only the instrumentation should be modified.

The procedure for model-based evaluation of a combination zone partition - instrumentation is helpful to determine the best combination to perform Single Inhabitant Location Tracking. Moreover, these criteria are computed offline and do not depend on the behavior of the inhabitant. If the inhabitant changes, there is no need to perform the evaluation again and if the instrumentation or the topology of the house changes, no new learning phase is required to evaluate this new configuration.

However, the quantitative criteria could not be used only in order to perform automatically an optimization of the choice of a combination zone partition - instrumentation, because it will surely lead to a partition of one zone and an instrumentation of one sensor observing this zone. Thus, an expert is required to read the indications given by the evaluation procedure (unlocationable zones, inaccurate locations) in order to improve the combination zone partition - instrumentation.

### 4.4.2. Illustration on a case study

#### First combination zone partition - instrumentation

The whole approach of evaluation-aided design of Smart Home is illustrated in a case study. Its combination zone partition - instrumentation is shown in Fig. 4.6. The house without its instrumentation is exactly the same as in Fig. 2.3 (a), it is composed of an open space including dining room (H), kitchen (I) and living room (J); a corridor (B); toilets (G); a shower (F); two bedrooms (A and C) and a bathroom (D and E). The major difference with the previously considered case study is that the bathroom is split into two zones, this has a real importance, since the location of the inhabitant in the bathtub is critical and thus strongly required.

Concerning the instrumentation, in a first time 5 motion detectors ($MD_1$ in the first bedroom, $MD_2$ in the corridor, $MD_3$ in the second bedroom, $MD_4$ in the bathroom and $MD_5$ in the open space) and two door barrier sensors ($DB_1$ placed on the door between the corridor and the toilets and $DB_2$ placed on the door between the corridor and the bedroom 1) are considered. The aim of this case study is to illustrate how the designer may take advantage from the evaluation criteria in order to improve his chosen combination zone partition - instrumentation. The improvement loop is illustrated for Single Inhabitant Location Tracking only, in order to show all the possibilities of improvement. The improvement procedure has also to be done for multiple inhabitants once a satisfying combination for single inhabitant has been defined. The improvement loop for multiple inhabitants is not shown because it is similar to the procedure for single inhabitant.

This first example of combination zone partition - instrumentation is named $(P^1, I^{(1,1)})$.

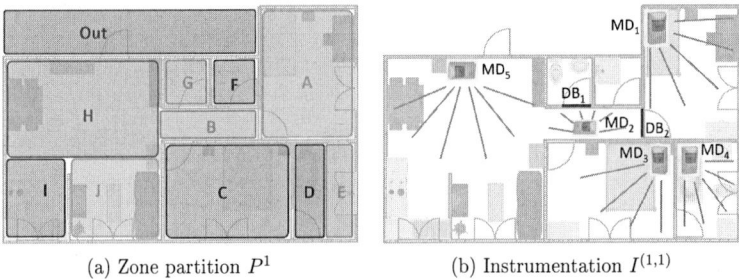

(a) Zone partition $P^1$        (b) Instrumentation $I^{(1,1)}$

Figure 4.6.: First envisaged combination zone partition - instrumentation for the case study of evaluation-aided improvement

The procedure for iterative model building and evaluation of Algorithm. 4.1 is applied. The results are the following.

The unlocationable zones are $Q_{UZ}^{(1,1)} = \{F; Out\}$ and $|Q_{UZ}^{(1,1)}| = 2$. This result shows that the inhabitant will never be located in zone $F$ or in zone $Out$. This is particularly critical for $Out$ since the location of the inhabitant in this zone is meaning that the inhabitant is not at home. This information is mandatory for some applications, for instance applications aiming to detect health problems when the inhabitant is at home. These applications should be deactivated each time the inhabitant is outside in order to avoid false alarm. Thus, the ability to give the information "inhabitant is out of the home" is required. This combination $(P^1, I^{(1,1)})$ does not have the ability to guarantee that the apartment is empty.

Moreover for this combination $(P^1, I^{(1,1)})$, the accurate locations of a single inhabitant are $Q_{a_1}^{(1,1)} = \{A, B, C, D, H\}$ and the inaccurate locations of a single inhabitant $Q_{i_1}^{(1,1)} = \{(A, B, C, D, E, F, G, H, I, J, Out), (H, I, J), (B, G), (A, B), (D, E)\}$. This leads to the following results concerning the accurate-location-ability. Strong 1-accurate-location-ability is not obtained because there is at least one transition between $Q_{a_1}^{(1,1)}$ and $Q_{i_1}^{(1,1)}$ in $Est(DMA^{(1,1)})$ (for example the transition $\delta_{Est}^{1,1}(B, DB_2\_1) = (A, B)$). Thus there exists a sequence of events such that the estimated location becomes inaccurate after it has been accurate. For instance the sequence of events $MD_2\_1, DB_2\_1$ leads to the sequence of estimated locations $(A, B, C, D, E, F, G, H, I, J, Out), B, (A, B)$ where $B$ is accurate and the first and third estimated locations are inaccurate.

However, since $Q_{a_1}^{(1,1)}$ is not empty and there exist at least one loop on $Q_{a_1}^{(1,1)}$ (for instance the self-loop on state $A$, formally written $\delta_{Est}^{(1,1)}(A, MD_1\_1) = A$), the weak 1-accurate-location-ability is guaranteed. At least one sequence of events exists such that the location becomes accurate and all the subsequent locations are accurate. For instance the sequence $DB_2\_1, MD_1\_1, MD_1\_1^*$ leads the location to be inaccurate in $(A, B, C, D, E, F, G, H, I, J, Out)$, then inaccurate in $(A, B)$, finally accurate in $A$ and it remains accurate in $A$ since $\delta_{Est}^{(1,1)}(A, MD_1\_1) = A$.

**Second combination zone partition - instrumentation**

Based on this complete analytical evaluation for single inhabitant, it is possible to propose a new combination zone partition - instrumentation and to evaluate it. In order to solve the problem of the two unlocationables zones ($F$ and $Out$), a new combination $(P^1, I^{(2,1)})$ with additional sensors is proposed (see Fig. 4.7). A new sensor $MD_6$ (motion detector) is added in the shower (zone $F$). This sensor is detecting motion only in zone $F$. A new sensor $DB_3$ (door barrier sensor) is added on the front door between the dining room and outside. This sensor is detecting motion in zones $H$ and $Out$.

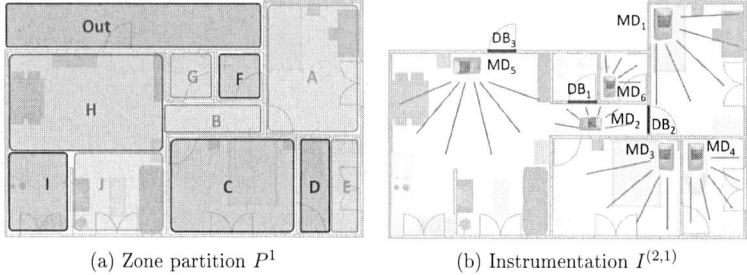

(a) Zone partition $P^1$          (b) Instrumentation $I^{(2,1)}$

Figure 4.7.: Second envisaged combination zone partition - instrumentation for the case study of evaluation-aided improvement

Using the same criteria, the couple $(P^1, I^{(2,1)})$ is evaluated. The set of unlocatable zones has been reduced to the emptyset $Q_{UZ}^{(2,1)} = \emptyset$, meaning that the motion of an inhabitant is always observable by a sensor. The sets of accurate and inaccurate states are computed: $Q_{a_1}^{(2,1)} = \{A, B, C, D, F, H\}$ and $Q_{i_1}^{(2,1)} = \{(A, B, C, D, E, F, G, H, I, J, Out), (H, Out),$ $(H, I, J), (B, G), (A, B), (D, E)\}$. There are still 5 inaccurate estimations as previously. In addition, there is still no Strong 1-accurate-location-ability but Weak 1-accurate-location-ability. Adding two sensors led to a reduced number of unlocatable zones but did not improve the accuracy of location.

### Third combination zone partition - instrumentation

In a third step, the zone partition is modified and $P^2$ is considered (see Fig. 4.8). Since neither the zone $I$ nor the zone $J$ are in $Q_{a_1}^{(2,1)}$ and since they are always together in the inaccurate locations of $Q_{i_1}^{(2,1)}$, these two zones are merged and named $I$. In addition, a new instrumentation $I^{(3,2)}$ is considered (see Fig. 4.8). The sensor $DB_2$ (door barrier) is removed from the door between the corridor (zone $B$) and the first bedroom (zone $A$) because $A \in Q_{a_1}^{(2,1)}$ and $B \in Q_{a_1}^{(2,1)}$ and $(A, B) \in Q_{a_1}^{(2,1)}$. Moreover, the sensor $DB_1$ (door barrier) is removed from the door between the corridor (zone $B$) and the toilets (zone $G$) and replaced by a motion detector $MD_7$ in the toilets and observing only the zone $G$. Finally the placement of the sensor $MD_4$ is modified so that it observed only the zone $D$ and a new sensor $MD_8$ is added in the bathroom, observing only the zone $E$.

Note that the situation of zones $I$ and $J$ was exactly the same as the situation of zones $D$ and $E$. However, there was a strong constraint on keeping the zone $D$ in the bathroom, thus leading to an impossible merging of the zones $D$ and $E$. Consequently, the two solutions are illustrated. Either merge the two zones (like $I$ and $J$) when an accurate location of the inhabitant in one of the two zones is not strongly required, or modify the position of the existing sensors and potentially add a new one (like $D$ and $E$ where the placement of sensor $MD_4$ is modified and an additional sensor $MD_8$ is placed in zone $E$) when the accurate location of the inhabitant in one of the zones (in this case, in zone $D$) is strongly required.

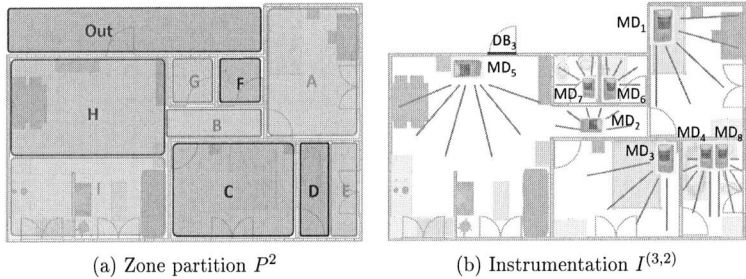

(a) Zone partition $P^2$  (b) Instrumentation $I^{(3,2)}$

Figure 4.8.: Third envisaged combination zone partition - instrumentation for the case study of evaluation-aided improvement

Once again, the combination $(P^2, I^{(3,2)})$ is evaluated. $Q_{UZ}^{(3,2)}$ is still the empty set. The set of accurate estimations is now $Q_{a_1}^{(3,2)} = \{A, B, C, D, E, F, G, H\}$ and the set of inaccurate estimations $Q_{i_1}^{(3,2)} = \{(A, B, C, D, E, F, G, H, I, J, Out), (H, Out), (H, I)\}$ has decreased from three elements $(A, B)$, $(B, G)$ and $(D, E)$ and the previously inaccurate estimation $(H, I, J)$ is now reduced to $(H, I)$. Still there is no Strong 1-Accurate-Location-Ability but only Weak 1-Accurate-Location-Ability. In conclusion, this last combination shows better performances than the two previous ones; however this is not yet optimal for Single Inhabitant Location Tracking.

In addition, to this analytical performance evaluation, the simulation-based evaluation has also been performed for single inhabitant for this combination. A 5-minute long scenario has been simulated and the different criteria (*accuracy*, *precision* and *recall*) have been calculated along this time.

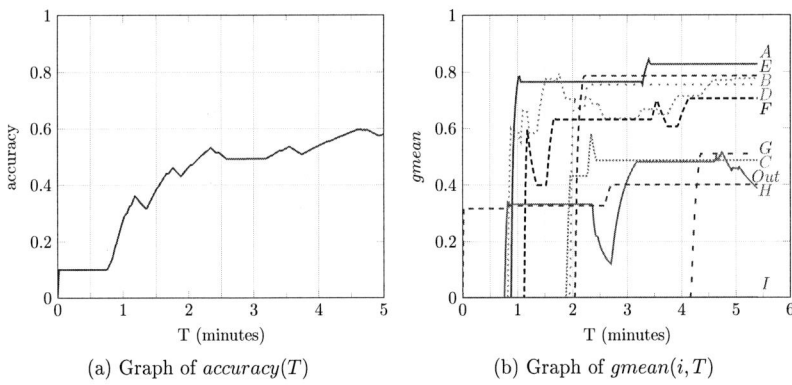

(a) Graph of $accuracy(T)$  (b) Graph of $gmean(i, T)$

Figure 4.9.: Evolution of the overall *accuracy* and of the local performances using $gmean(i, T)$ along the time $T$ for the third combination

The evolution of the *accuracy* along the time is shown in Fig. 4.9 (a). At the end of the scenario, the *accuracy* has a value of 55%.

Moreover, in order to have a closer loop on the causes of this low *accuracy*, the evolution of the *gmean* criterion is shown in Fig. 4.9 (b). At the end of the scenario, this criterion is equal to 0 for the zone $I$ and is quite low for the zones $G$, $C$, $Out$ and $H$ (between 38% and 51%).

Based on this simulation and on the poor evaluated performance, it has been tried to add new sensors in the home, thus leading to the fourth considered combination zone partition - instrumentation below.

**Fourth combination zone partition - instrumentation**

For this fourth combination, the same zone partition $P^2$ is considered and new sensors are added in the house, leading to a new instrumentation $I^{(4,2)}$. This combination $(P^2, I^{(4,2)})$ is shown in Fig. 4.10. 6 floor pressure sensors are now considered in addition to the previous sensors ($FP_1$ just at the entrance of the house, $FP_2$ in the living room just before the corridor, $FP_3$ at the entrance of the first bedroom, $FP_4$ at the entrance of the second bedroom, $FP_5$ and $FP_6$ on both sides of the door between the second bedroom and the bathroom).

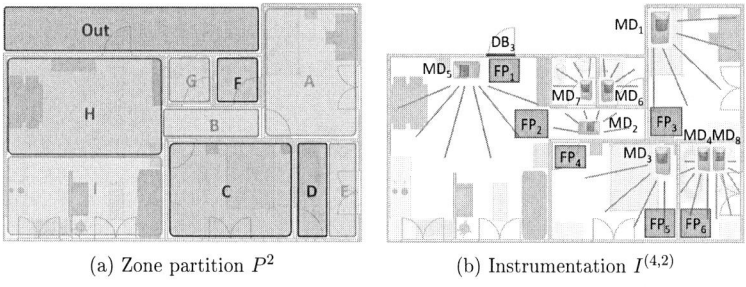

(a) Zone partition $P^2$        (b) Instrumentation $I^{(4,2)}$

Figure 4.10.: Fourth envisaged combination zone partition - instrumentation for the case study of evaluation-aided improvement

The analytical evaluation leads to the same performance indicators as the previous one (Fig. 4.8). In order to compare these last two combinations, the simulation of the exact same scenario as previously described was also performed with this new combination.

The evolution of the *accuracy* along the time is shown in Fig. 4.11 (a). On the same graph, the *accuracy* during the same scenario for the previous combination (the third one) is also shown in order to compare the combinations. At the end of the scenario, the *accuracy* has a value of 65% which is an increase of 10% compared to the previous combination.

Moreover, in order to have a closer loop on the different zones and to see in details the effect of the new sensors, the evolution of the *gmean* criterion is shown in Fig. 4.11 (b). At the end of the scenario, this criterion is now equal to 40% for the zone $I$, this is due to the sensor $FP_2$ improving the detection of the inhabitant in the zone $H$ and when an inhabitant is detected in this zone $H$, if the next event to come is the rising edge of $MD_5$, then the inhabitant is estimated as being either in $H$ or in $I$, thus improving the location of the inhabitant in $I$. This new sensor $FP_2$ has also an impact on the *gmean* value for the zone $H$ increasing from 38% to 65%. Another improvement concerns the zone $C$ where the two new sensors $FP_4$ and $FP_5$ lead to an increase from 48% to 75%. There is no change for the value of $G$ and $Out$ because these two zones were not concerned at all by the new instrumentation.

To conclude this comparative study of four combinations, note that the analytical evaluation allows proposing a first improvement of the combination but is then limited (there is

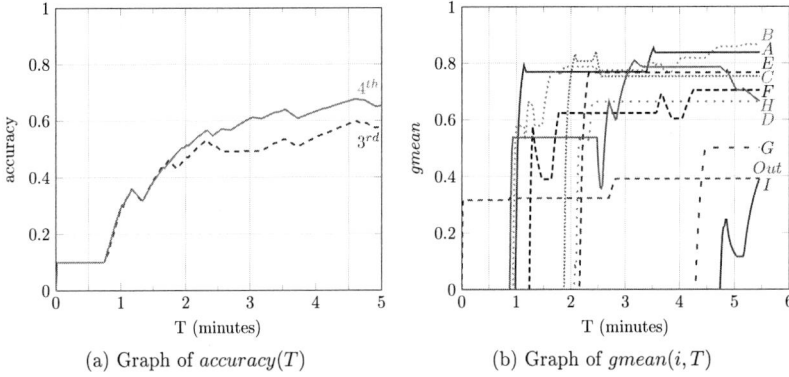

(a) Graph of $accuracy(T)$        (b) Graph of $gmean(i, T)$

Figure 4.11.: Evolution of the overall *accuracy* (compared with the third instrumentation) and of the local performances using $gmean(i, T)$ along the time $T$ for the fourth combination

no difference in terms of analytical performances between the third and the fourth combinations). Then, the simulation-based evaluation allowed highlighting the weakness of the third combination and the advantages of the fourth combination.

At the end of these two loops of evaluation-aided improvement, the combination zone partition -instrumentation seems to be well adapted for Single Inhabitant Location Tracking. The two same improvement loops should then be done for Multiple Inhabitants Location Tracking. Note that improving the combination for a single inhabitant already led to an improved combination for multiple inhabitants since $N_{max}$ was equal to 4 for the first combination and is now equal to 6 for the fourth combination.

## Conclusion

In this chapter, two approaches to evaluate the relevance of a combination zone partition - instrumentation have been introduced. The analytical evaluation provides a guarantee of the ability to accurately track the location of the inhabitants and the maximal number of inhabitants being trackable by a given combination. It also provides the designer with candidates to improve the combination (unlocationable zones for instance). The simulation-based evaluation allows testing particular or critical scenarios. It provides the designer with quantitative criteria values of the accuracy of location (globally and locally) and candidates to improve the combination zone partition - instrumentation. Based on these two evaluation approaches, a whole evaluation-aided improvement method has been proposed and illustrated on a detailed case study. The details on the implementation of the Smart Home emulator and its use for simulation-based evaluation are presented in the Appendix of this thesis.

In the next chapter of the thesis, the contributions of the thesis are summarized and an outlook for future work is given.

# Conclusion

## Summary

The objective of the present thesis was to propose a contribution in the field of Ambient Assisted Living and Smart Homes using the paradigms, theory and tools of the Discrete Event Systems domain. Considering an inhabited Smart Home as a spontaneous discrete event generator (based on several assumptions on the instrumentation of the home and on the behavior of the inhabitants) allowed proposing a new approach for indoor Location Tracking. This new approach is based on an explicit model and can be applied in the case of single inhabitant, *a priori* defined and constant number of inhabitants (greater than one) or *a priori* unknown number of inhabitants. In order to develop this approach, three main contributions were proposed, detailed and illustrated in this thesis.

### A Procedure for modeling the detectable motion of the inhabitants

In a first time, a definition of the model of detectable motion of the inhabitants in a Smart Home has been given and a detailed procedure allowing any designer (i.e. possibly with no competence in Discrete Event System theory) to model the detectable motion of $N$ inhabitants ($N \in \mathbb{N}^*$) has been proposed. This procedure is based on a basic description of the zones of interest of the considered Smart Home and of its instrumentation. Several modeling approaches have been compared on a similar case study in order to evaluate the impact of the choices in the formalization on the resulting model. Furthermore, the impact of the granularity of the model has been discussed. Finally, the whole approach has been widely illustrated on real case studies.

### Several algorithms for Location Tracking

In a second step, several algorithms for model-based Location Tracking were proposed. The different cases of single, multiple or *a priori* unknown number of inhabitants have been envisaged. These algorithms have been detailed and illustrated on scenarios of motion and action of inhabitants in their Smart Home. Finally, the possibility to relax the assumption of fault-free sensors, thus considering potential sensor faults, has been discussed. Existing approaches for model-based fault detection and isolation from the field of DES have been applied and the possibility to perform fault-tolerant Location Tracking has been envisaged. However, the conclusion on this topic leads to maintain the assumption of fault-free sensors at the current point of development.

### A global method for evaluation and improvement of the designer's choices

Finally, a method allowing evaluating the relevance for Location Tracking of the expert's choices while designing a Smart Home has been proposed. This contribution is divided in three main topics. First, an analytical evaluation procedure has been detailed. This procedure allows giving guarantees on the Location Tracking ability of a given Smart Home. This phase of evaluation is performed offline and is only based on the model. In a second time, a simulation-based evaluation is proposed. This dynamic evaluation is based on new criteria allowing evaluating the accuracy of location for particular scenarios. This approach has been proved to be complementary to the analytical one. Finally, a procedure based on an improvement loop has been developed. Using the two different evaluation procedures and their criteria, this improvement loop allows the expert to design a Smart Home with guarantees on its Location Tracking ability and an idea of the final result.

Another contribution, in addition to the three main ones, has been proposed regarding the implementation of the approach. During this thesis, a computer program combining a Smart Home emulator, a Location Tracking simulator and a performance evaluator has been implemented (see the appendix of the thesis). The algorithms for systematic generation of the Detectable Motion Automaton have also been implemented in another program. These two programs can be used in order to apply the whole approach of model-based Location Tracking proposed in this thesis.

# Outlook

A complete approach for Location Tracking is proposed in this thesis. However, there is still outlook for future work toward improving this approach or other challenges of the fields of AAL and Smart Homes that may be faced using DES techniques.

### Short term outlook

A first outlook is related to the assumption of non-distinguishable inhabitants, when dealing with multiple inhabitants. A perspective is to relax this assumption, either by using particular sensor, detecting only particular inhabitants (pets for instance), that would potentially be wearable sensors, or just by using different models of the detectable motion for the different inhabitants but keeping the same set of sensors for both inhabitants.

Another outlook consists in really developing a new approach for sensor Fault Detection and Isolation dedicated to Smart Homes. Based on this approach, it may be possible to propose a fault-tolerant Location Tracking method and thus relax the assumption of fault-free sensor. This approach will surely lead to new evaluation criteria in order to provide an FDI guarantee (diagnosability) and a fault-tolerant Location Tracking guarantee (for instance by evaluating the Location Tracking performances with $S$-1 sensors, i.e. with one sensor missing).

Finally, it seems that the DES theory may be fruitful to contribute to the field of ADL recognition. Finite Automata may be used for the modeling of different ADL and a Finite Automaton player may perform ADL recognition, in quite the same way a Finite Automaton player performs Location Tracking in this thesis.

### Long term outlook

In the long term, the smart control of the Smart Home actuators could be envisaged. A pro-active Smart Home may benefit from the DES approaches, particularly from the Supervisory Control Theory.

Finally, all these approaches will have to be tested in real environments. Their aim is of course to be deployed in real Smart Homes. Collaborations with industrial partners must be accentuated in order to transfer these theoretical results.

# Extended summaries in French and German

## Résumé en langue française

### Introduction

L'espérance de vie a augmenté de manière continue dans les dernières décennies et devrait continuer à croître dans les prochaines années. Cette augmentation entraîne de nouveaux défis concernant l'autonomie et l'indépendance des personnes âgées. Le développement de maisons intelligentes est une piste pour répondre à ces défis et permettre aux personnes de vivre plus longtemps dans un environnement sûr et confortable. Rendre une maison intelligente consiste à y installer des capteurs, des actionneurs et un contrôleur de manière à pouvoir observer le comportement de ses habitants et agir sur leur environnement (voir FIG. S.1).

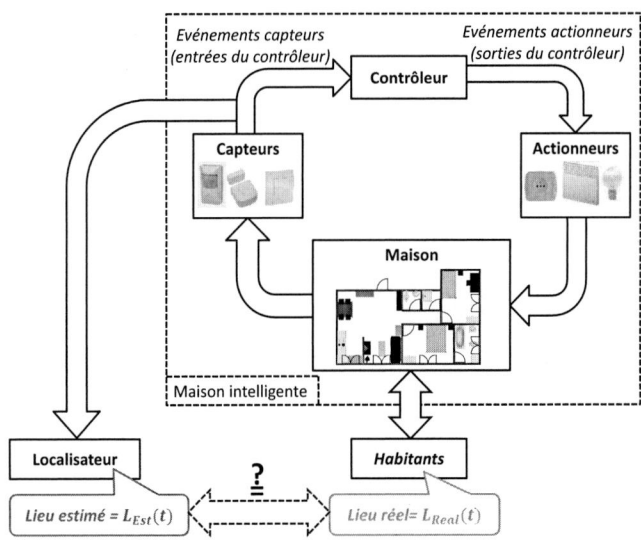

FIGURE S.1.: Concept de la maison intelligente et de la localisation en ligne de ses habitants

Les principaux domaines de recherche dans ce domaine sont relatifs à l'amélioration du confort, de la sécurité et de la santé. La plupart de ces approches s'appuient sur la localisation en temps réel des habitants dans leur maison. Localiser les habitants en temps réel consiste à estimer leur lieu à tout moment, à partir de l'observation des valeurs des différents capteurs (voir FIG. S.1). Dans cette thèse, une nouvelle approche complète permettant la localisation d'un nombre *a priori* inconnu d'habitants basée sur le modèle est proposée. Cette approche tire parti des paradigmes, de la théorie et des outils des Systèmes à Événements Discrets.

Une vue d'ensemble de la méthode développée et des contributions de cette thèse est donnée sur la FIG. S.2. L'utilisation d'automates à états finis pour modéliser le mouvement détectable des habitants ainsi que des méthodes permettant de construire ce modèle ont été développées (étapes 1 et 2). A partir de ce modèle, plusieurs algorithmes permettant de localiser de manière efficace les habitants ont été définis (étape 3). Enfin, plusieurs approches pour l'évaluation des performances de l'instrumentation d'une maison intelligente pour un objectif de localisation ont été proposées (étape 4 et 6). A l'aide des procédures d'évaluation développées, une méthode complète permettant l'amélioration itérative de l'instrumentation d'une maison intelligente a été mise en place (boucle 2, 4 et 5, boucle 2, 6 et 7 et étape 8). Les détails relatifs à ces différentes contributions sont donnés dans les trois paragraphes suivants.

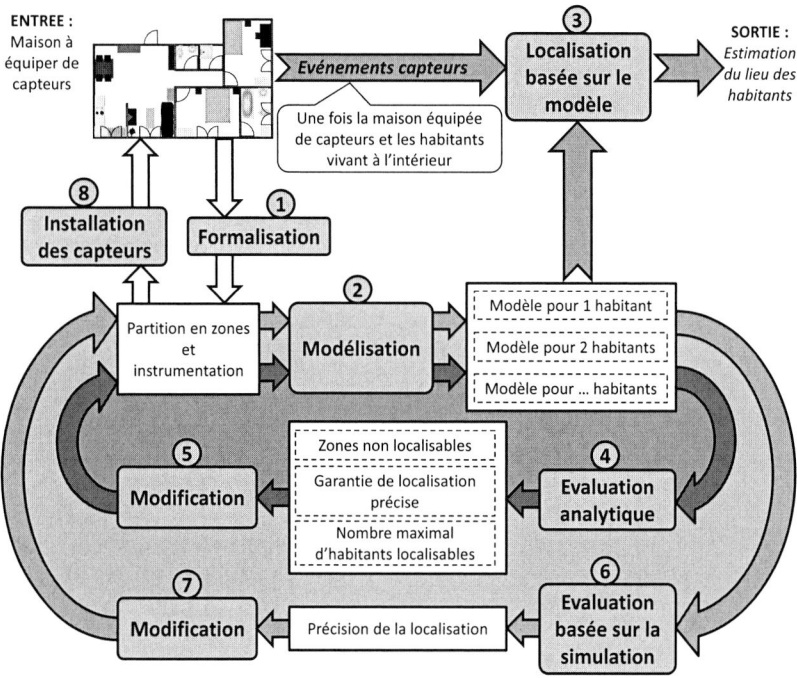

FIGURE S.2.: Vue d'ensemble des contributions de la thèse

## Modélisation du mouvement détectable

La première étape de l'approche proposée consiste à construire un modèle de la mobilité détectable d'un ou plusieurs habitants dans leur maison intelligente. Il ne s'agit pas d'un modèle du comportement de l'habitant car un habitant se comporte de manière non déterministe, ni même stochastique, il agit de façon arbitraire et potentiellement irrationnelle. Par conséquent, le modèle proposé doit être robuste à tous les comportements possibles des habitants. Dans un premier temps, un modèle du mouvement détectable d'un seul habitant doit être construit. Dans un second temps, le précédent modèle sert de base à la construction d'un modèle du mouvement détectable de $N$ personnes dans l'habitat.

Le modèle pour un habitant seul proposé dans cette thèse est appelé Automate du Mouvement Détectable ($DMA$). Il s'agit d'un automate à états finis où :

- Chaque état représente une zone de l'habitat. Une zone peut être une ou plusieurs pièces (salle de bain, chambre) ou une partie d'une pièce (autour du four dans la cuisine, la baignoire dans la salle de bain). L'environnement (l'habitat ainsi que l'extérieur) doit être divisé en zones qui ne se superposent pas, ces dernières étant définies par un expert selon l'objectif final de la localisation.

- Les événements de l'automate sont les fronts montants et descendants des différents capteurs installés dans l'habitat. Seuls des capteurs binaires (détecteurs de mouvements, tapis sensitifs, capteurs barrière sur les portes) sont considérés dans l'approche développée. Dans l'automate, chaque transition représente le mouvement de l'habitant d'une zone à une autre (ou à l'intérieur d'une zone dans le cas des boucles, ayant le même état source et état destination).

- La fonction de transition est éventuellement non déterministe (pour un même état source et un même événement, plusieurs états destination peuvent être définis), ceci permet de représenter un capteur observant plusieurs zones.

- Enfin, tous les états de l'automate sont initiaux, cela permet de représenter la non-connaissance du lieu dans lequel se trouve initialement l'habitant.

Plusieurs approches sont proposées dans la thèse (mais non détaillées ici) pour construire systématiquement un tel modèle à partir d'une formalisation plus ou moins détaillée d'une partition en zones de l'habitat et de son instrumentation. Quelle que soit la méthode utilisée, le résultat est toujours un $DMA$ respectant la description énoncée ci-dessus. Un exemple est donné en FIG. S.3, la topologie du cas d'étude est donné en FIG. S.3 (a), son instrumentation en FIG. S.3 (b), la partition en zones choisie en FIG. S.3 (c) et le modèle obtenu pour un habitant seul en FIG. S.3 (d).

A partir du $DMA$, une méthode permettant d'obtenir automatiquement un modèle du mouvement détectable de $N$ personnes, appelé $MIDMA_N^{red}$ est proposé. Il s'agit une fois encore d'un automate à états finis. Chaque état représente le nombre de personnes dans chacune des zones ; les événements sont les fronts montant et descendants des capteurs ; chaque transition représente le mouvement d'une ou plusieurs personnes d'une zone vers une autre (ou à l'intérieur d'une zone) ; enfin chaque état de l'automate est initial car le lieu initial de chacun des habitants est inconnu.

## Localisation des habitants en temps réel

A partir des automates $DMA$ et $MIDMA_N^{red}$ $\forall N \geq 2$, des algorithmes permettant de localiser en temps réel un nombre fixe d'habitants ou un nombre a priori inconnu et variable d'habitants sont proposés. Concernant la localisation d'un nombre fixe d'habitants (y compris le cas d'un habitant seul), il suffit de "jouer" l'automate correspondant avec les événements capteurs observés lors des mouvements des habitants. Le ou les états actifs de l'automate représentent alors une estimation plus ou moins précise du lieu des habitants à chaque instant. Lorsque le nombre d'habitants est a priori inconnu et variable, tous les modèles sont utilisés. Lorsqu'un comportement observé par les capteurs n'est pas reproductible par le modèle de $i$ habitants,

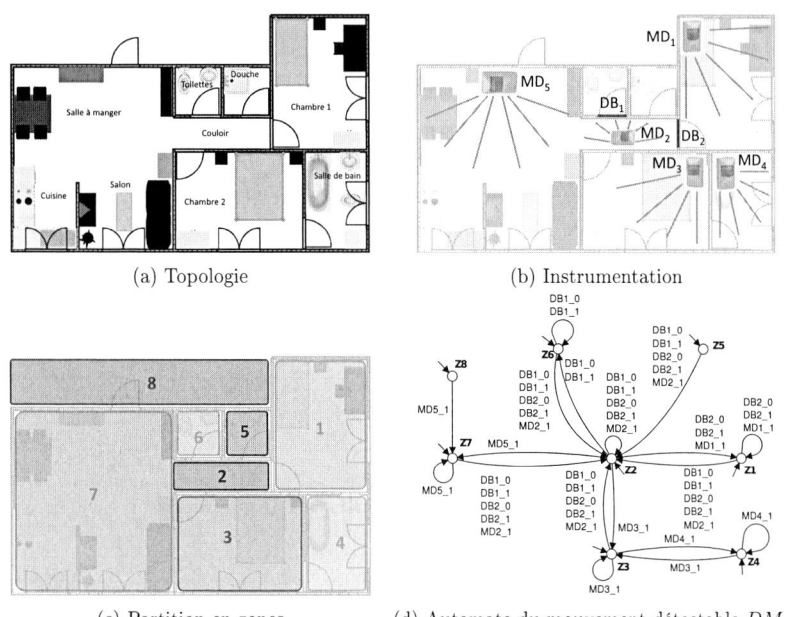

(a) Topologie · · · · · · · · · · · · · · · · · · · · (b) Instrumentation

(c) Partition en zones · · · · · · · · · · · · · (d) Automate du mouvement détectable $DMA$

FIGURE S.3.: Illustration de la modélisation sur un cas d'étude

on en déduit qu'il y a au moins $i+1$ habitants dans l'appartement. En effet, les capteurs sont supposés ne pas avoir de comportement fautif et en conséquence, le comportement observé de $i$ habitants ou moins est supposé être toujours reproductible par le modèle de $i$ habitants. S'il ne l'est pas, cela signifie qu'il y a plus de $i$ habitants. Lors de la thèse, l'impact d'éventuelles fautes capteurs (émission d'un événement inattendu ou non émission d'un événement attendu) a également été étudié et une approche pour la détection et la localisation de ces fautes a été proposée, elle n'est pas détaillée dans ce résumé.

### Evaluation des performances pour la localisation et application à l'amélioration de l'instrumentation

Les modèles utilisés pour la localisation des habitants sont très dépendants du choix d'une combinaison Partition en zones - Instrumentation. Par conséquent, le résultat de l'algorithme de localisation dépend également de ce choix de combinaison. Deux procédures permettant d'évaluer la pertinence d'une combinaison Partition en zones - Instrumentation pour la localisation sont proposées.

La première approche permet d'évaluer analytiquement les performances de localisation via trois critères : calcul de l'ensemble des "zones non localisables" ; caractérisation de la "garantie de localisation précise" à partir d'une formulation forte et d'une formulation faible. Le dernier critère analytique est relatif au "nombre maximal d'habitants localisables". Tous ces critères sont calculés à partir des modèles et sont donc indépendants des habitants considérés et de leur comportement.

La seconde approche, basée sur la simulation de comportements humains, permet de proposer des critères complémentaires. Les modèles et l'algorithme de localisation sont testés pour certains scénarios critiques. Cette évaluation est effectuée en comparant le lieu estimé des habitants (donné par l'algorithme de localisation) avec leur lieu réel (supposé connu). Cette comparaison est formalisée à l'aide d'une "matrice de confusion". A partir de cette matrice, un critère global appelé "précision globale" permet d'évaluer la proportion du temps durant lequel le lieu est correctement estimé. En plus de ce critère global, des critères locaux pour chaque zone peuvent être calculés : il s'agit de la "précision" et du "rappel". Ils permettent d'évaluer pour chaque zone, respectivement, la proportion du temps durant lequel le lieu estimé représente bien le lieu réel et la proportion du temps durant lequel le lieu réel est bien estimé. Ces deux critères sont combinés à l'aide de leur moyenne géométrique pour obtenir un critère unique pour chaque zone.

Les deux approches d'évaluation sont complémentaires et peuvent être utilisées dans une boucle d'amélioration du choix de la combinaison Partition en zones - Instrumentation d'un habitat intelligent. La partition en zones ne doit être modifiée qu'en dernier recours car elle dépend fortement de l'utilisation du résultat de la localisation (détection des problèmes de santé, amélioration du confort) et peut donc comporter des zones critiques relatives à cette utilisation. Cependant, les résultats tant analytiques (zones non localisables, garantie de localisation précise, nombre maximal d'habitants localisables) que basés sur la simulation (précision globale, précision et rappel locaux) donnent des indications au concepteur pour améliorer la combinaison en agissant sur son choix d'instrumentation. Plusieurs boucles d'évaluation et de modification de la combinaison peuvent être nécessaires avant d'obtenir une combinaison et les modèles associés satisfaisants pour la localisation en temps réel des habitants.

## Conclusion

Cette thèse propose :

- une approche complète pour la modélisation du mouvement détectable des habitants dans un habitat intelligent,

- l'utilisation de ces modèles pour localiser en temps réel un nombre *a priori* inconnu et variable d'habitants,

- une procédure pour l'évaluation des choix du concepteur concernant l'instrumentation dans un objectif de localisation.

Les différentes contributions proposées ont été implémentées. Notamment, un émulateur de maison intelligente, couplé à un simulateur de l'algorithme de localisation et à un évaluateur de performances, permet de tester une ou plusieurs instrumentations pour certains scénarios critiques en immergeant l'utilisateur dans un habitat émulé via une manette de jeu vidéo.

Les perspectives restent nombreuses. Dans ces travaux, les capteurs sont supposés fonctionner de manière idéale et surtout sans fautes. Les fautes capteurs pourraient probablement être prises en compte et le concept de localisation tolérante aux fautes développé. Le formalisme automate à états finis représente également une piste prometteuse pour contribuer à la reconnaissance des activités de la vie quotidienne (ADL) des habitants. Ce formalisme permet aussi d'envisager le contrôle d'une maison proactive (i.e. comprenant également des actionneurs) en utilisant des techniques de la théorie du contrôle par supervision (SCT) par exemple.

## Kurzfassung in deutscher Sprache

### Einführung

In den meisten Industrieländern ist die Lebenserwartung in den letzten Jahrzehnten fortlaufend gestiegen und wird höchstwahrscheinlich noch weiter steigen. Dieser Anstieg führt zu neuen Herausforderungen hinsichtlich der Autonomie und Unabhängigkeit von älteren Menschen. Die Entwicklung von intelligenten Wohnungen ist ein Weg diesen Herausforderungen zu begegnen und es den Menschen zu ermöglichen, länger in einer sicheren und komfortablen Umgebung zu leben. Dazu stattet man solche Wohnungen mit Sensoren, Aktoren sowie einem Controller aus. Dies ermöglicht es, in Abhängigkeit vom Verhalten der Bewohner, die Wohnumgebung so zu beeinflussen, dass sich Sicherheit, Gesundheit und Komfort verbessern (siehe Abb. S.4).

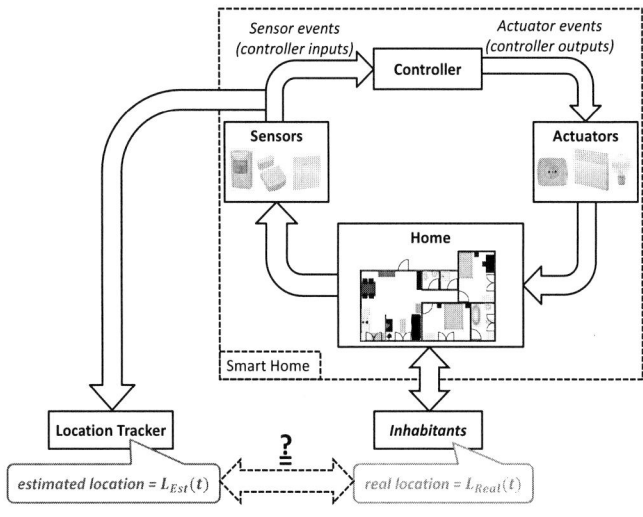

Abbildung S.4.: Konzept der intelligenten Wohnung und der Echtzeitlokalisierung von Bewohnern

Ansätze, die dies zum Ziel haben, basieren meistens auf Methoden, die es ermöglichen Menschen innerhalb ihrer Wohnung in Echtzeit zu lokalisieren (siehe Abb. S.4). In dieser Dissertation wird ein neuer Ansatz für eine modellbasierte Lokalisierung einer *a priori* unbekannten Anzahl von Bewohnern vorgestellt. Dieser Ansatz basiert auf der Theorie, den Paradigmen und den Werkzeugen aus dem Gebiet der ereignisdiskreten Systeme.

Ein Überblick über den Ansatz und den wissenschaftlichen Beitrag dieser Dissertation wird auf Abb. S.5 gezeigt. Es werden endliche Automaten eingesetzt, um die von den Sensoren erfassbaren Bewohnerbewegungen zu modellieren. Verschiedene Verfahren zur Erzeugung solcher Automaten werden angewendet (Schritte 1 und 2). Basierend auf diesen Modellen werden Algorithmen definiert, mittels derer die Bewohner wirksam lokalisiert werden können (Schritt 3). Abschließend werden Methoden vorgeschlagen, die dazu dienen die Relevanz der Sensorinstrumentierung für die Lokalisierung zu bewerten (Schritte 4 und 6). Mit Hilfe dieser Bewertungsverfahren wird eine vollständige Methode zur iterativen Verbesserung der Sensorinstrumentierung einer intelligenten Wohnung vorgeschlagen (Schleife 2, 4 und 5, Schleife 2, 6 und 7 und Schritt 8). Diese Kontributionen werden in den nächsten Absätzen detailliert.

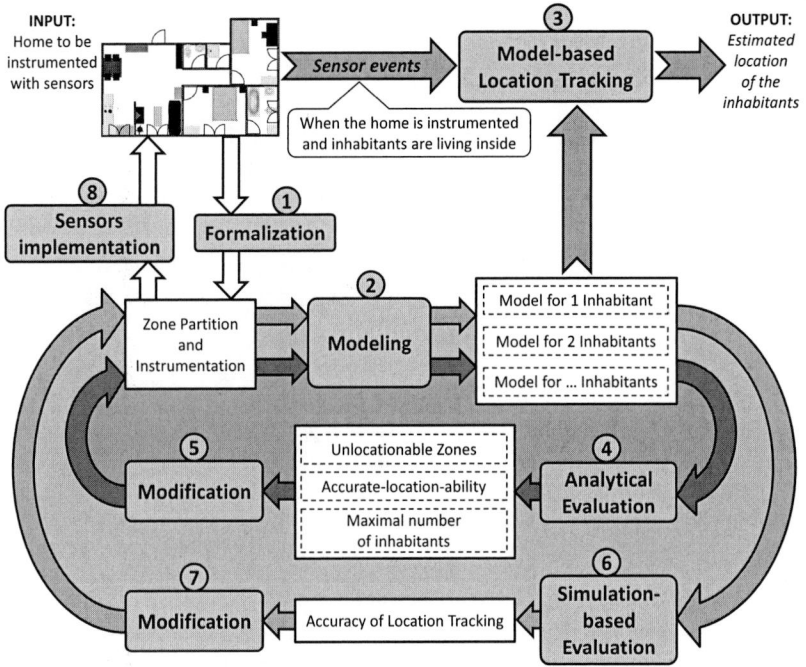

Abbildung S.5.: Wissenschaftlicher Beitrag der Dissertation

## Modellierung der erfassbaren Bewohnerbewegungen

Der erste Schritt des entwickelten Ansatzes besteht darin, ein Modell der erfassbaren Bewegungen eines oder mehrerer Bewohner in ihrer intelligenten Wohnung zu erstellen. Es handelt sich nicht um ein Modell des Bewohnerverhaltens sondern um ein Modell, das robust gegen jedes mögliche Bewohnerverhalten sein muss. Tatsächlich wird angenommen, dass das Bewohnerverhalten weder deterministisch noch stochastisch ist. Es ist arbiträr und möglicherweise irrational. Zunächst soll ein Modell der erfassbaren Bewegungen eines einzigen Bewohners erstellt werden. Anschließend wird dieses Modell als Basis dienen, um das Modell der erfassbaren Bewegungen von $N$ Bewohnern zu erstellen.

Das vorgeschlagene Modell für einen einzige Bewohner heißt "Automat der erfassbaren Bewegungen" ($DMA$). Es handelt sich um einen endlichen Automaten,

- dessen Zustände sogenannte Zonen darstellen. Eine Zone kann entweder ein oder mehrere Zimmer (Badezimmer, Schlafzimmer) oder ein Teil eines Zimmers (die Badewanne im Badezimmer, die Umgebung des Backofens in der Küche) sein. Die ganze Umgebung (die Wohnung sowie deren unmittelbare Umgebung) soll mit nicht-überlappenden Zonen bedeckt werden. Die Zonen werden von einem Experten je nach Ziel der Lokalisierungsaufgabe bestimmt.

- Die Ereignisse des Automaten sind die steigenden und fallenden Flanken der Sensoren, die in der Wohnung installiert werden. Es werden nur binäre Sensoren (Bewegungsmelder, sensitiven Matten, Tür Barriere Sensoren) berücksichtigt.

- Jeder Übergang im Automaten stellt die Bewegung des Bewohners von einer Zone nach einer anderen Zone dar (oder in einer Zone im Fall von Schleifen, wenn der Ursprungszustand und der Zielzustand der Gleiche sind). Die Übergangsfunktion ist gegebenenfalls nicht-deterministisch (für einen gegebenen Ursprungszustand und ein gegebenes Ereignis können mehrere Zielzustände bestimmt werden). Damit kann ein Sensor dargestellt werden, der mehr als nur eine Zone beobachtet.

- Schließlich sind alle Zustände initial, denn der initiale Ort des Bewohners ist nicht bekannt.

Mehrere Modellierungsmethoden werden in der Dissertation vorgeschlagen, jede basiert auf einer mehr oder weniger formalisierten Aufteilung der Wohnung in Zonen und ihrer Sensorinstrumentierung. Unabhängig von der Modellierungsmethode ist das Ergebnis immer ein $DMA$, der die oben stehende Beschreibung erfüllt. Ein Fallbeispiel wird in Abb. S.6 gegeben, die Topologie wird in Abb. S.6 (a) gezeigt, die Sensorinstrumentierung in Abb. S.6 (b), die Partition in Zonen in Abb. S.6 (c) und das Modell für einen Einzelbewohner in Abb. S.6 (d).

Eine Methode basierend auf der $DMA$ wird vorgeschlagen, um ein Modell der erfassbaren Bewegungen von N Bewohnern zu erstellen. Dieses Modell ist $MIDMA_N^{red}$ benannt. Es handelt sich dabei wiederum um einen endlichen Automat. Jeder Zustand stellt die Anzahl von Bewohnern in jeder Zone dar. Die Ereignisse sind die steigenden und fallenden Flanken der Sensoren. Jeder Übergang stellt die Bewegung von einem oder mehreren Bewohnern von einer Zone nach einer anderen Zone (oder in einer Zone) dar. Schließlich sind alle Zustände des Automaten initial, denn der initiale Ort jedes Bewohners ist nicht bekannt.

## Echtzeitlokalisierung von der Bewohnern

Basierend auf den Automaten $DMA$ und $MIDMA_N^{red}$ werden Algorithmen vorgeschlagen, die die Echtzeitlokalisierung von einer konstanten Anzahl oder von einer a priori unbekannten und variablen Anzahl von Bewohnern erlauben. Die Lokalisierung von einer konstanten Anzahl von Bewohnern (einschließlich des Falls von einem Einzelbewohner) kann folgendermaßen realisiert werden: der dazugehörige Automat wird mit den beobachteten Sensorereignissen "gespielt". Der oder die aktiven Zustände des Automaten stellen eine mehr oder weniger präzise Abschätzung des gegenwärtigen Orts jedes Bewohners dar. Falls die Anzahl von Bewohnern a priori unbekannt und variabel ist, werden alle Modelle genutzt. Wenn das von den Sensoren beobachtete Verhalten nicht reproduzierbar mit dem Modell für $i$ Bewohnern ist, dann bedeutet es, dass mindestens $i$ Bewohnern anwesend in der Wohnung sind. Wenn angenommen wird, dass die Sensoren immer fehlerfrei arbeiten, wird das beobachtete Verhalten von $i$ oder weniger Bewohnern immer reproduzierbar mit dem Modell für $i$ Bewohnern sein. Wenn es nicht reproduzierbar ist, bedeutet es, dass mehr als i Bewohnern anwesend sind. Im Rahmen der Dissertation wird darüber hinaus der Einfluss von Sensorfehlern (senden eines unerwarteten Signals oder nicht-Sendung eines erwarteten Signals) untersucht und ein Ansatz für Fehlererkennung und -identifizierung vorgeschlagen.

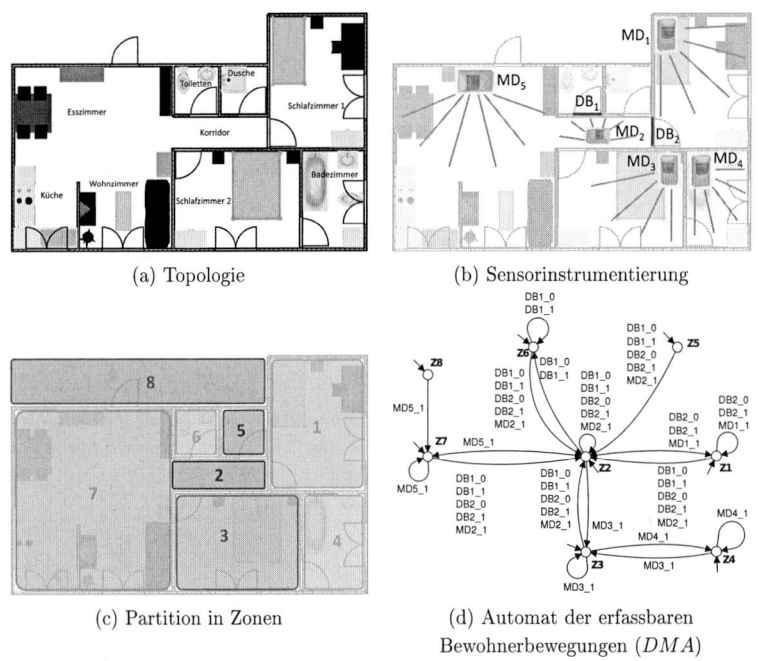

(a) Topologie

(b) Sensorinstrumentierung

(c) Partition in Zonen

(d) Automat der erfassbaren
Bewohnerbewegungen ($DMA$)

Abbildung S.6.: Beispiel-Modellierung

## Bewertung der Relevanz einer Sensorinstrumentierung für die Lokalisierung und Anwendung zur Verbesserung der Sensorinstrumentierung

Die Modelle, die für die Lokalisierung von den Bewohnern genutzt werden, sind sehr abhängig von der Kombination "Partition in Zonen" - "Sensorinstrumentierung". Daher ist das Ergebnis des Lokalisierungsalgorithmus abhängig von dieser Kombination. Zwei Verfahren werden vorgeschlagen, um die Relevanz einer Kombination "Partition in Zonen" - "Sensorinstrumentierung" für die Lokalisierung zu bewerten.

Das erste Verfahren ist analytisch und erlaubt die Angabe von Garantien für das Lokalisierungsergebnis. Zum Beispiel wird die Menge von "Unlokalisierbaren Zonen" bewertet und es werden Kriterien über die "Garantie der präzisen Lokalisierung" bestimmt. Eine starke sowie eine schwache Formulierung dieser Garantie werden gegeben. Das letzte analytische Kriterium besteht in der "maximalen Anzahl lokalisierbarer Bewohner". Alle diese Kriterien basieren rein auf den Modellen und sind folglich unabhängig von konkreten Bewohnern und deren Verhalten.

Zusätzlich zu diesen Kriterien wird ein Ansatz basierend auf der Simulation von menschlichem Verhalten vorgeschlagen. Dieser Ansatz führt zu neuen Kriterien und einem zweiten Bewertungsverfahren. Die Grundidee dafür besteht darin, dass kritische Szenarien simuliert werden können, um die Algorithmen und die Modellen zu prüfen. Die Bewertung besteht im Vergleich des geschätzten Orts (Ergebnis des Lokalisierungsalgorithmus) mit dem wirklichen Ort der Bewohnern (bekannt durch die Simulation). Dieser Vergleich wird mittels einer so genannte "Wahrheitsmatrix" geleistet. Aus dieser Matrix wird als Kriterium die "globale

113

Genauigkeit" bestimmt. Dieses Kriterium erlaubt es, den zeitlichen Anteil der korrekten Bewohnerlokalisierung zu quantifizieren. Neben diesem Kriterium können lokalen Kriterien für jede Zone bewertet werden. Es handelt um die "Genauigkeit" (precision) und die "Trefferquote" (recall). Die Genauigkeit ist der Teil der Zeit, während dessen der geschätzte Ort korrekt ist. Die Trefferquote ist der Teil der Zeit, während dessen der tatsächliche Ort korrekt geschätzt wird. Diese zwei Kriterien können durch einen geometrischen Mittelwert kombiniert werden, um ein Einzelkriterium für jede Zone zu erhalten.

Die zwei Bewertungsansätze sind komplementär und können iterativ genutzt werden, um die Kombination "Partition in Zonen" - "Sensorinstrumentierung" einer intelligenten Wohnung zu verbessern. Die Partition in Zonen sollte nur als letztes Mittel geändert werden, da sie in der Regel übergeordneten Zielen folgend definiert wird (Erkennung von Gesundheitsproblemen, Verbesserung des Komforts). Allerdings liefern die analytischen Ergebnisse (Unlokalisierbare Zonen, Garantie der präzisen Lokalisierung, maximale Anzahl von lokalisierbaren Bewohnern) sowie simulationsbasierte Ergebnisse (globale Genauigkeit, lokale Genauigkeit und Trefferquote) dem Experten Hinweise, mit denen er die Sensorinstrumentierung verbessern kann. Prinzipiell können mehrere Iterationen aus Bewertung und Änderung erforderlich sein, bevor die Kombination und die Modelle für die Echtzeitlokalisierung von den Bewohnern geeignet sind.

## Zusammenfassung und Ausblick

In dieser Dissertation wird ein vollständiger Ansatz vorgeschlagen, um die erfassbaren Bewohnerbewegungen in einer intelligenten Wohnung zu modellieren. Die Modelle werden zur Echtzeitlokalisierung von einer a priori unbekannten und variablen Anzahl von Bewohnern genutzt und die Sensorinstrumentierung im Hinblick auf die Lokalisierung bewertet. Die verschiedenen Verfahren wurden implementiert. Insbesondere wurden eine Emulator einer "intelligenten Wohnung", ein Lokalisierungssimulator und ein Leistungsbewerter implementiert. Damit können eine oder verschiedene Sensorinstrumentierungen für kritische Szenarien überprüft werden, indem man die Szenarien in der emulierten Wohnung mittels Joystick vorgibt.

Es bieten sich zahlreiche Anknüpfungspunkte für weitere Arbeiten. In dieser Arbeit wird vorausgesetzt, dass die Sensoren fehlerfrei arbeiten. In zukünftigen Arbeiten können Sensorfehler berücksichtigt werden, um die entwickelten Methoden zu robustifizieren. Der Formalismus der endlichen Automaten stellt außerdem in einem interessanten Weg dar, um die Methoden zur Erkennung der Alltagsaktivitäten von den Bewohnern (ADL) zu verbessern. Der automatenbasierte Formalismus könnte auch genutzt werden, um systematisch eine Ansteuerung der Aktuatoren einer intelligenten Wohnung zu entwickeln (zum Beispiel mittels die "Supervisory Control Theory").

# Bibliography

AAL JP (2013). Objectives of the Ambient Assisted Living joint programme. http://www.aal-europe.eu/about/objectives/ (as of 26 Sep 2013).

Abdul Majid, M. (2011). *Human Behavior Modelling: an investigation using traditional disrete event and combined discrete event and agent-based simulation.* PhD thesis.

Abidine, B. M. and Fergani, B. (2012). A Comparative Study of Four Classifiers for Activity Recognition in Smart Homes. In *Proceedings of the Première Conférence Nationale sur les Télécommunications, CNT'2012, At Guelma, Algeria.*

Allègre, W., Burger, T., Berruet, P., and Antoine, J.-Y. (2012). A Non-Intrusive Monitoring System for Ambient Assisted Living Service Delivery. In *Proceedings of the 10th International Conference on Smart Homes and Health Telematics, ICOST'12, Artimino, Italy, Lecture Notes in Computer Science*, volume 7251, pages 148–156.

Anliker, U., Ward, J., Lukowicz, P., Tröster, G., Dolveck, F., Baer, M., Keita, F., Schenker, E., Catarsi, F., Coluccini, L., Belardinelli, A., Shklarski, D., Alon, M., Hirt, E., Schmid, R., and Vuskovic, M. (2004). AMON: A Wearable Multiparameter Medical Monitoring and Alert System. *IEEE Transactions on Information Technology in Biomedicine*, 8(4):415–427.

ANSI/IEEE100 (1997). *The IEEE Standard Dictionary of Electrical and Electronics Terms according to ANSI/IEEE standard 100-1988.* IEEE Standards Office, New York.

Bahl, P. and Padmanabhan, V. (2000). RADAR: An Inbuilding RF-based user location and tracking system. In *Proceedings of IEEE INFOCOM*, volume 2, pages 775–784.

Beringer, R., Sixsmith, A., Campo, M., Brown, J., and McCloskey, R. (2011). The "Acceptance" of Ambient Assisted Living: Developing an Alternate Methodology to This Limited Research Lens. In *Proceedings of the 9th International Conference on Smart Homes and Health Telematics, ICOST'11, Montreal, Canada, Lecture Notes in Computer Science*, volume 6719, pages 161–167.

Borges, I., Sinclair, D., Mollenkopf, H., Rayner, P., Bond, R., and Parent, A.-S. (2008). Older people an Information and Communication Technologies - An Ethical approach. http://www.age-platform.eu/images/stories/EN/pdf_AGE-ethic_A4-final-2.pdf (as of 26 Sep 2013).

Boyer, J. P., Tan, K., and Gunter, C. A. (2006). Privacy Sensitive Location Information Systems in Smart Buildings. In *Security in Pervasive Computing, Lecture Notes in Computer Science*, volume 3934, pages 149–164.

Cardinaux, F., Bhowmik, D., Abhayaratne, C., and Hawley, M. S. (2011). Video based technology for ambient assisted living: a review of the litterature. *Journal of Ambient Intelligence and Smart Environments*, 3(3):253–269.

Cassandras, C. and Lafortune, S. (2009). *Introduction to Discrete Event Systems*. Springer-verlag edition.

Chen, C., Zhang, D., Sun, L., Hariz, M., and Yuan, Y. (2012). Does Location Help Daily Activity Recognition. In *Proceedings of the 10th International Conference on Smart Homes and Health Telematics, ICOST'12, Artimino, Italy, Lecture Notes in Computer Science*, volume 7251, pages 83–90.

Chikhaoui, B., Wang, S., and Pigot, H. (2011). Activity Recognition in Smart Environments: An Information Retrieval Problem. In *Proceedings of the 9th International Conference on Smart Homes and Health Telematics, ICOST'11, Montreal, Canada, Lecture Notes in Computer Science*, volume 6719, pages 33–40.

Cook, D. J. (2006). Health Monitoring and Assistance to Support Aging in Place. *Journal of Universal Computer Science*, 12(1):15–29.

Cook, D. J. and Das, S. K. (2007). How Smart are our Environments? An Updated Look at the State of the Art. *Journal of Pervasive and Mobile Computing*, 3(2):53–73.

Danancher, M., Roth, M., Lesage, J.-J., and Litz, L. (2011). A comparative study of three model-based FDI approaches for Discrete Event Systems. In *Proceedings of the 3rd International Workshop on Dependable Control of Discrete Systems, DCDS'11, Saarbrücken, Germany*, pages 29–34.

Danancher, M., Lesage, J.-J., and Litz, L. (2012). Indoor Location Tracking Based on a Discrete Event Model. In *Proceedings of the 10th International Conference on Smart Homes and Health Telematics, ICOST'12, Artimino, Italy, Lecture Notes in Computer Science*, volume 7251, pages 262–265.

Danancher, M., Faraut, G., Lesage, J.-J., and Litz, L. (2013a). A DES Simulator for Location Tracking of Inhabitants in Smart Home. In *Proceedings of the 8th EUROSIM Congress on Modelling and Simulation, Cardiff, United Kingdom*, pages 330–335.

Danancher, M., Lesage, J.-J., Litz, L., and Faraut, G. (2013b). A Discrete Event Model for Multiple Inhabitants Location Tracking. In *Proceedings of the 9th IEEE International Conference on Automation Science and Engineering, CASE'13, Madison, United States*, pages 922–927.

Danancher, M., Lesage, J.-J., Litz, L., and Faraut, G. (2013c). Online Location Tracking of a Single Inhabitant based on a State Estimator. In *Proceedings of the IEEE Conference on Systems, Man and Cybernetics, SMC'13, Manchester, United Kingdom*, pages 391–396.

Das, B., Chen, C., Seelye, A. M., and Cook, D. J. (2011). An Automated Prompting System for Smart Environments. In *Proceedings of the 9th International Conference on Smart Homes and Health Telematics, ICOST'11, Montreal, Canada, Lecture Notes in Computer Science*, volume 6719, pages 9–16.

Das, S. K., Roy, N., and Roy, A. (2006). Context-aware resource management in multi-inhabitant smart homes: A framework based on Nash H-learning. *Pervasive and Mobile Computing*, 2:372–404.

Eurostat (2010). Eurostat Population Projections 2010-based (Europop2010).

Fleury, A., Vacher, M., and Noury, N. (2010). SCM-Based Multimodal Classification of Activities of Daily Living in Health Smart Homes: Sensors, Algorithms, and First Experimental results. *IEEE Transactions on Information Technology in Biomedicine*, 14(2):274–283.

Floeck, M. (2010). *Activity Monitoring and Automatic Alarm Generation in AAL-enabled Homes*. PhD thesis, University of Kaiserslautern.

Floeck, M. and Litz, L. (2009). Inactivity Patterns and Alarm Generation in Senior Citizens' Houses. *Proceedings of the European Control Conference, ECC'09, Budapest, Hungary*, pages 3725–3730.

Floeck, M., Litz, L., and Rodner, T. (2011). An Ambient Approach to Emergency Detection based on Location Tracking. In *Proceedings of the 9th International Conference on Smart Homes and Health Telematics, ICOST'11, Montreal, Canada, Lecture Notes in Computer Science*, volume 6719, pages 296–302.

Floeck, M., Litz, L., and Spellerberg, A. (2012). *Monitoring Patterns of Inactivity in the Home with Domotics Networks*, pages 258–282. IOS press amsterdam edition.

Grauel, J. and Spellerberg, A. (2008). Attitudes and Requirements of Elderly People Towards Assisted Living Solutions. *Constructing Ambient Intelligence, Communications in Computer and Information Science*, 11:197–206.

Hightower, J. and Borriello, G. (2001). A Survey and Taxonomy of Location Systems for Ubiquitous Computing. Technical report, University of Washington, Computer Science and Engineering.

Jalal, A., Uddin, M. Z., Kim, J. T., and Kim, T.-S. (2011). Daily Human Activity Recognition Using Depth Silhouettes and R Transformation for Smart Home. In *Proceedings of the 9th International Conference on Smart Homes and Health Telematics, ICOST'11, Montreal, Canada, Lecture Notes in Computer Science*, volume 6719, pages 25–32.

Kautz, H., Fox, D., Etzioni, O., Borriello, G., and Arnstein, L. (2002). An Overview of the Assisted Cognition Project. *AAAI Technical Report WS-02-02*, pages 60–65.

Kleinberger, T., Becker, M., Ras, E., Holzinger, A., and Müller, P. (2007). Ambient intelligence in assisted living: enable elderly people to handle future interfaces. In *Proceedings of the 4th International Conference on Universal access in human-computer interaction, Beijing, China*, pages 103–112.

Kubat, M., Holte, R., and Matwin, S. (1998). Machine Learning for the Detection of Oil Spills in Satellite Radar Images. In *Machine Learning*, pages 195–215.

Lankri, S., Berruet, P., and Philipe, J.-L. (2009). Multi-Level Reconfiguration in the DANAH Assistive System. In *IEEE International Conference on Systems, Man, and Cybernetics, SMC'09, San Antonio, United States*, pages 1084–1089.

Lankri, S., Berruet, P., Rossi, A., and Philipe, J.-L. (2008). Architecture and Models of the DANAH Assistive System. *Proceedings of the 3rd International Workshop on Services Integration in Pervasive Environments, SIPE'08, New York, United States*, pages 19–24.

Liao, L., Fox, D., and Kautz, H. (2005). Location-Based Activity Recognition using Relational Markov Networks. In *Proceedings of the Nineteenth International Conference on Artificial Intelligence, IJCAI'05, San Francisco, United States*, pages 773–778.

Lu, C.-H. and Fu, L.-C. (2009). Robust Location-Aware Activity Recognition Using Wireless Sensor Network in an Attentive Home. *IEEE Transactions on Automation Science and Engineering*, 6(4):598–609.

Makonin, S. and Popowich, F. (2011). An Intelligent Agent for Determining Home Occupancy Using Power Monitors and Light Sensors. In *Proceedings of the 9th International Conference on Smart Homes and Health Telematics, ICOST'11, Montreal, Canada, Lecture Notes in Computer Science*, volume 6719, pages 236–240.

McKeever, S., Ye, J., Coyle, L., Dobson, S., and Bleakley, C. (2010). Activity Recognition using Temporal Evidence theory. *Journal of Ambient Intelligence and Smart Environments*, 2(3):253–269.

Nazerfard, E., Rashidi, P., and Cook, D. J. (2011). Using Association Rule Mining to Discover Temporal Relations of Daily Activities. In *Proceedings of the 9th International Conference on Smart Homes and Health Telematics, ICOST'11, Montreal, Canada, Lecture Notes in Computer Science*, volume 6719, pages 49–56.

Noury, N., Fleury, A., Rumeau, P., Bourke, A., Laighin, G., Rialle, V., and Lundy, J.-E. (2007). Fall detection - Principle and Methods. *Proceedings of the 29th Annual International IEEE EMBS Conference, Lyon, France*, pages 1663–1666.

Noury, N., Hadidi, T., Laila, M., Fleury, A., Villemazet, C., Rialle, V., and Franco, A. (2008). Level of Activity, Night and Day Alternation, and well being measured in a Smart Hospital Suite. *Proceedings of the 30th Annual International IEEE EMBS Conference, Vancouver, Canada*, pages 3328–3331.

Noury, N., Quach, K. A., Berenguer, M., Teyssier, H., Bouzid, M.-J., Goldstein, L., and Giordani, M. (2009). Remote Follow Up of Health Through the monitoring of Electrical Activities on the Residential Power Line - Preliminary Results of an experimentation. In *Proceedings of the 11th IEEE International Conference of E-health networking, application and services, Healthcom'2009, Sydney, Australia*, pages 9–13.

Noury, N., Quach, K. A., Berenguer, M., Teyssier, H., Bouzid, M.-J., Goldstein, L., and Giordani, M. (2010). Ubiquitous but non invasive evaluation of the activity of a person from a unique index built on the electrical activities on the residential power line. *Proceedings of the 12th IEEE International Conference of E-health networking, application and services, Healthcom'2010, Lyon, France*, pages 1–6.

Pandalai, D. and Holloway, L. (2000). Templates languages for fault monitoring of timed discrete event processes. *IEEE Transactions on Automatic Control*, 45:868–882.

Park, H., Park, T., and Son, S. H. (2013). A Comparative Study of Privacy Protection Methods for Smart Home Environments. *International Journal of Smart Home*, 7(2).

Pollack, M. E., Brown, L., Colbry, D., McCarthy, C. E., Orosz, C., Peintner, B., Ramakrishnan, S., and Tsamardinos, I. (2003). Autominder: an intelligent cognitive orthotic system for people with memory impairment. *Robotics and Autonomous Systems*, 44:273–282.

Poujaud, J., Noury, N., and Lundy, J.-E. (2008). Identification of inactivity behavior in Smart Home. In *Proceedings of the 30th Annual International IEEE EMBS Conference, Vancouver, Canada*, pages 2075–2078.

Priyantha, N. B., Chakraborty, A., and Balakrishnan, H. (2000). The Cricket Location-Support System. In *Proceedings of MOBICOM 2000*, pages 32–43. ACM Press.

Rahal, Y., Pigot, H., and Mabilleau, P. (2008). Location Estimation in a Smart Home: System Implementation and Evaluation Using Experimental Data. *International Journal of Telemedicine and Applications*, article No. 4.

Robles, R. J. and Kim, T.-h. (2010). Applications, Systems and Methods in Smart Home Technology: A Review. *International Journal of Advanced Science and Technology*, 15:37–47.

Roth, M. (2010). *Identification and Fault Diagnosis of Industrial Closed-Loop Discrete Event Systems*. PhD thesis, Ecole Normale Supérieure de Cachan and University of Kaiserslautern.

Roth, M., Lesage, J.-J., and Litz, L. (2009). A residual inspired approach for fault localization in DES. In *Proceedings of the 2nd IFAC Workshop on Dependable Control of Discrete Event Systems, DCDS'09, Bari, Italy*, pages 347–352.

Rougier, C., Auvinet, E., Rousseau, J., Mignotte, M., and Meunier, J. (2011). Fall Detection from Depth Map Video Sequences. In *Proceedings of the 9th International Conference on Smart Homes and Health Telematics, ICOST'11, Montreal, Canada, Lecture Notes in Computer Science*, volume 6719, pages 121–128.

Roy, A., Das Bhaumik, S. K., Bhattacharya, A., Basu, K., Cook, D. J., and Das, S. K. (2003). Location Aware Resource Management in Smart Homes. In *Proceedings of the IEEE International Conference on Pervasive Computing and Communications, PerCom'03, San Diego, United States*, pages 481–488.

Sampath, M., Sengupta, R., Lafortune, S., Sinnamohideen, K., and Teneketzis, D. (1996). Failure diagnosis using discrete-event models. *IEEE Transactions on Control Systems Technology*, 4 (2):105–124.

Schulze, B., Floeck, M., and Litz, L. (2009). Concept and Design of a Video Monitoring System for Activity recognition and Fall Detection. In *Proceedings of the 7th International Conference on Smart Homes and Health Telematics, ICOST'09, Tours, France, Lecture Notes in Computer Science*, volume 5597, pages 182–189.

Sehili, M. A., Lecouteux, B., Vacher, M., Portet, F., Istrate, D., Dorizzi, B., and Boudy, J. (2012). Sound Environment Analysis in Smart Home. In *Proceedings of the 3rd International Joint Conference on Ambient Intelligence, AmI'12, Pisa, Italy*, pages 208–223.

Seth Long, S. and Holder, L. B. (2011). Using Graph to Improve Activity Prediction in Smart Environment Based on Motion Sensor Data. In *Proceedings of the 9th International Conference on Smart Homes and Health Telematics, ICOST'11, Montreal, Canada, Lecture Notes in Computer Science*, volume 6719, pages 57–64.

Shu, S. and Lin, F. (2007). Detectability of Discrete Event Systems. *IEEE Transactions on Automatic Control*, 52(12):2356–2359.

Shu, S. and Lin, F. (2013). I-Detectability of Discrete-Event Systems. *IEEE Transactions on Automation Science and Engineering*, 10(1):187–196.

Skubic, M., Alexander, G., Popescu, M., Rantz, M., and Keller, J. (2009). A Smart home application to eldercare: Current status and lessons learned. *Technology and Health Care*, 17:183–201.

Stanley, R. P. (2012). *Enumerative Combinatorics*, volume 1, second edition. Cambridge University Press edition.

Storf, H., Kleinberger, T., Becker, M., Schmitt, M., Bomarius, F., and Prueckner, S. (2009). An Event-Driven Approach to Activity Recognition in Ambient Assisted Living. In *Proceedings of the European Conference on Ambient Intelligence, AmI'09, Salzburg, Austria*, pages 123–132.

Tan, B. L. (2007). A Study to model human behavior in discrete event simulation (DES) using Simkit. Master's thesis.

Vacher, M., Portet, F., Fleury, A., and Noury, N. (2010). Challenges in the Processing of Audio Channels for Ambient Assisted Living. *12th IEEE International Conference on E-Health Networking, Application and Services, Healthcom'10, Lyon, France*, pages 330–338.

van Glabbeek, R. and Ploeger, B. (2008). Five Determinisation Algorithms. In *Proceedings of the 13th International Conference on Implementation and Applications of Automata, CIAA'08, San Fransisco, United States*, pages 161–170.

van Kasteren, T., Englebienne, G., and Kröse, B. (2010). Activity recognition using semi-Markov models on real world smart home datasets. *Journal of Ambient Intelligence and Smart Environments*, 2:311–325.

Wilson, D. and Atkeson, C. (2005). Simultaneous tracking & activity recognition (star) using many anonymous, binary sensors. *Pervasive Computing*, pages 62–79.

World Health Organization (2012). Good health adds life to years. In *Global brief for World Health Day 2012*, http://www.who.int/ageing/publications/whd2012_global_brief/en/index.html (as of 26 Sep 2013).

Yan, W., Weber, C., and Wermter, S. (2011). A Hybrid Probabilistic Neural Model for Person Tracking based on a Ceiling-mounted Camera. *Journal of Ambient Intelligence and Smart Environments*, 3(3):237–252.

Yu, C.-R., Wu, C.-L., Lu, C.-H., and Fu, L.-C. (2006). Human Localization via Multi-Cameras and Floor Sensors in Smart Home. In *IEEE International Conference on Systems, Man, and Cybernetics, SMC'06, Taipei, Taiwan*, pages 3822–3827.

Zhang, S., McCullagh, P., Nugent, C., Zheng, H., and Black, N. (2011). A Subarea Mapping Approach for Indoor Localization. In *Proceedings of the 9th International Conference on Smart Homes and Health Telematics, ICOST'11, Montreal, Canada, Lecture Notes in Computer Science*, volume 6719, pages 80–87.

# Appendix: Software implementation of the emulated Smart Home

## Introduction

In Section 4.3 and in (Danancher et al., 2013a), a performance evaluation approach based on simulation is proposed. Thus, a computer-program integrating a Smart Home emulator, a location tracker and a performance evaluator has been developed. It is aimed to be used for any possible Smart Home. The expert can use this program to design any Smart Home topology and instrumentation, to test it quickly and have a visual estimation of the final result in a real Smart Home, and finally, this can be used for simulation-based performance evaluation of a combination zone partition-instrumentation.

The Smart Home emulator, the location tracker and the performance evaluator have been developed using *Python 2.7*[1] and its object-oriented paradigms because this programing language is well appropriated to quickly develop proof-of-concept. In addition, the module *pygame*[2] was used in order to simulate the interaction of the inhabitants with their environment and to provide a graphical representation.

The usage of the emulator, tracker and evaluator is illustrated on a case study shown in Fig. A.1. It is a Smart Home composed of three rooms (an open space composed by the living room and the kitchen, the bedroom and the bathroom). This house is divided in 4 zones, one for each room and one for outside the house. In addition, 4 sensors are installed. One motion detector in each room ($MD_1$, $MD_2$ and $MD_3$) and one door barrier sensor on the front door ($DB$).

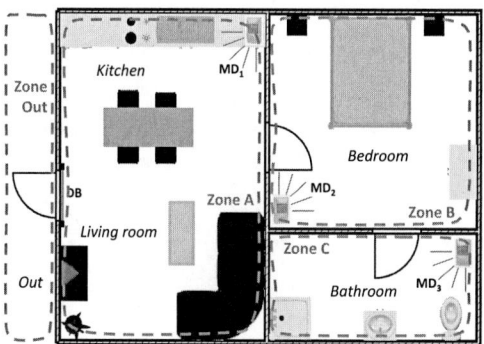

Figure A.1.: Case study of an emulated Smart Home

---

[1]http://www.python.org
[2]http://www.pygame.org

# I. Implementation of the Smart Home emulator

The Smart Home emulator allows representing a Smart Home in 2 dimensions (top view), the inhabitants moving in this environment, the different sensors and their value. For the considered case study, its representation when emulated through the program is shown in Fig. A.2. There is actually one inhabitant, but there can be several of them without problem. In this representation of the Smart Home, note that the walls are colored in black, the windows are colored in dark gray and the doors in pale gray. There are the three motion detectors, one in each room, and they are colored in green if detecting motion or in red if not detecting any motion. The same colors (red and green) hold for other kind of sensors (floor pressure sensors, door barrier sensors). These colors allow having a quick overview of the current value of all the sensors. For the considered case study of Fig. A.2, there is also the door barrier sensor on the front door of the house (not seen in the figure because the door in closed).

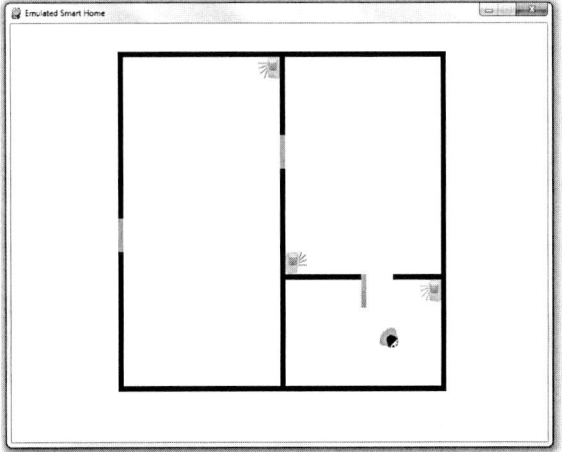

Figure A.2.: Screenshot of the Smart Home emulator

For a given apartment, its topology (walls, doors and windows) and its instrumentation (sensors) is described through an XML-file. For the considered case study, this XML-file is given in Fig. A.3. Each of the elements (wall, door, window or sensor) has a position (X and Y coordinates) and a size (along X and Y). Doors and windows have an predefined position "*opened*" and a predefined position "*closed*". Motion detectors have an angle of detection around a direction of detection and a range of detection. With this descriptive format, lots of Smart Homes can be defined although this format could be improved to consider other not so simple topologies.

This description of an instrumented Smart Home is the input of the Smart Home emulator which instantiates the different described objects (walls, doors, ...) by importing the XML-file in the code. The program is strongly object-oriented. A part of the UML Class Diagram is shown in Fig. A.4. Three main classes are defined:

- The class *Obstacle* defines the different possible obstacles composing the topology of the Smart Home like the walls. We defined the classes *Wall*, *Door* and *Window* which are

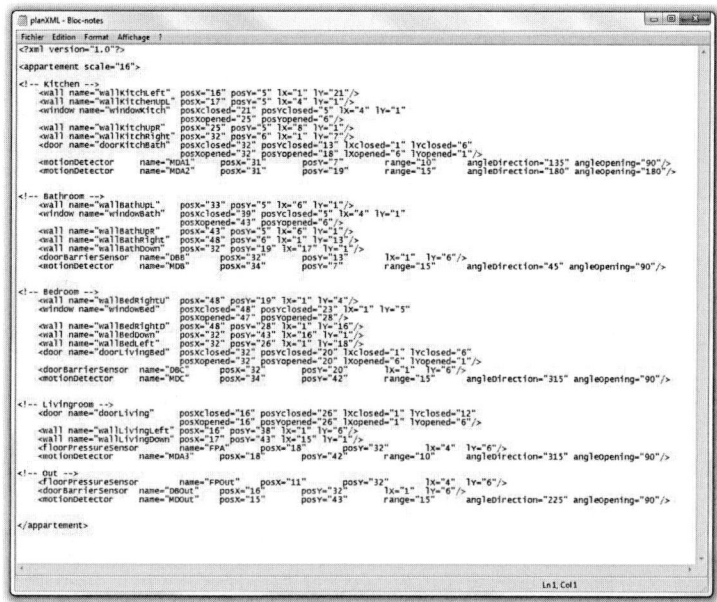

Figure A.3.: XML-file of the previously considered Smart Home (see Fig. A.2)

inheriting from the class *Obstacle*.

- The class *Sensor* defines the different possible sensors that can be used in the Smart Home. Since there are different types of sensors, additional classes inheriting from *Sensor* have been defined, for instance *Floor Sensor*, *Door Sensor* or *Motion Detector*. Their technology is defined through a specific model of functioning and specific attributes.

- The class *Inhabitant* defines the different inhabitants living in the Smart Home. Different kind of inhabitants can be defined, using the classes *Human* or *Pet* that both inherit from the class *Inhabitant*

In addition to these classes, the class *Occupied Smart Home* is defined. It is characterized by a list of instances of the class *Obstacle*, an instrumentation which is a list of instances of the class *Sensor* and a set of inhabitants which is a list of instances of the class *Inhabitant*. Finally, a queue of sensor events (FIFO type) is generated by the Smart Home emulator and used by the simulator.

It can also be noticed that the different classes are linked. The motions of the inhabitants are blocked by the different obstacles of the house, this is represented by the link between the classes *Inhabitant* and *Obstacle*. In a similar way, the sensors are reacting to the position and to the motion of the inhabitants; this is represented by the link between the classes *Sensor* and *Inhabitant*. Each inhabitant may also interact with the obstacles e.g. doors or windows can be opened only by humans (not by pets). Consequently, there is a link between the class *Human* and the class *Door* and a link between the class *Human* and the class *Window*. This allows taking into account different possible behaviors for different types of inhabitants.

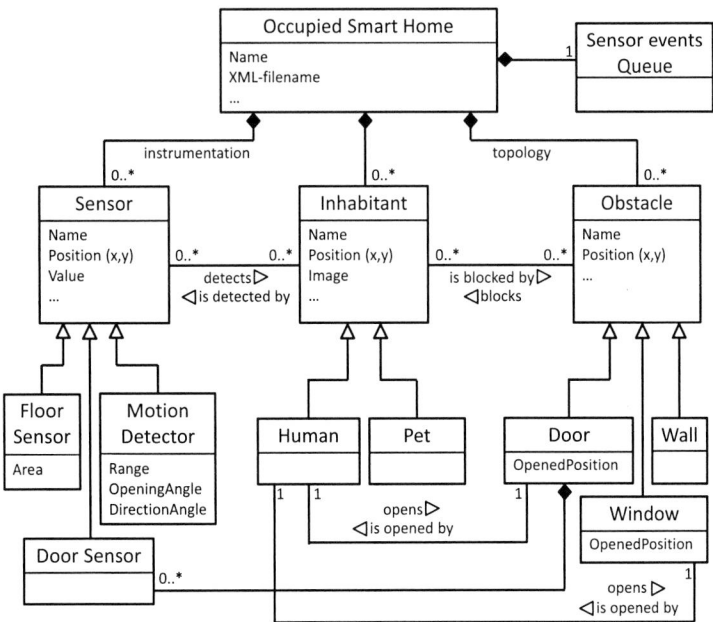

Figure A.4.: Part of the UML diagram of the Smart Home emulator

The user is immersed through a joystick or the computer keyboard in this emulated virtual smart environment reproducing the topology and the instrumentation of a real Smart Home. This allows avoiding implementing a model of the inhabitants' behavior. The behavior of each inhabitant is emulated by a real user, each playing the motion and action of an inhabitant in the house, constrained by the obstacles. Thus, the emulated behavior is assumed to be close to a real behavior of inhabitants in their Smart Home.

The sensors are reacting to the inhabitants' motion and action and provide the according sequence of events. Furthermore, in the emulator, the exact location of each inhabitant at each time is known. It is then possible to use the sequence of events online in order to perform Location Tracking (and thus to obtain the estimated location of the inhabitants at each time) and to evaluate the dynamic performance of the accuracy of the Location Tracking model by comparing the estimated location with the real one.

Moreover, using an emulated Smart Home allows guaranteeing the fault-free functioning of the sensors. This assumption is then satisfied and the algorithms strongly requiring it can be applied (Multiple Inhabitants Location Tracking for instance). However, it is also possible to introduce the faulty behavior of the sensors (sporadic lack of power supply, spurious signals, switch off delay of the motion detectors). Thanks to the implementation of the sensors using particular objects, different faults may be modeled and considered.

## II. Implementation of the online Location Tracking

The online Location Tracking is based on a model of the detectable motion of the inhabitants. The estimated location is given by the current state(s) of one of the model (among $DMA$, $MIDMA_2^{red}$, ..., obtained using the systematic generation approach of Chapter 2). Thus, the current estimated location is graphically represented by the according model and the according active state(s) colored in green (see Fig. A.5).

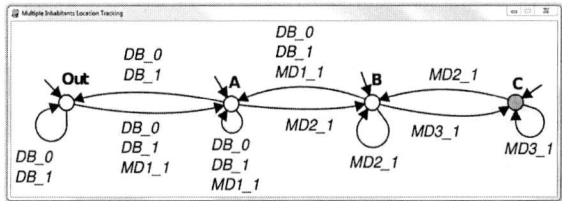

Figure A.5.: Screenshot of the online Location Tracking simulator

Note that the emulator and the location tracker are running in two parallel processes and moreover, their results (real location of the inhabitants in the Smart Home as in Fig. A.2 and estimated location as in Fig. A.5) are given in two different computer windows. Both are displayed simultaneously on the screen.

## III. Implementation of the online performance evaluator

The evaluation of performance is based on the comparison of the estimated location of the inhabitants with their real location. The estimated location is given by the location tracker whereas the real location is given by the emulator. The two are compared by the performance evaluator using the confusion matrix and criteria based on this matrix described in Chapter 4. This matrix and the criteria (*accuracy, precision, recall, gmean*) are displayed online in a third computer window. This window is shown in Fig. A.6.

| Performance Evaluator | | | | | | |
|---|---|---|---|---|---|---|
| accuracy = 0.815 | | | | | | |
| | | Estimated Location j | | | | |
| | | A | B | C | Out | recall(i) | gmean(i) |
| Real Location i | A | 0.153 | 0.014 | 0.000 | 0.006 | 0.887 | 0.749 |
| | B | 0.016 | 0.204 | 0.002 | 0.000 | 0.919 | 0.850 |
| | C | 0.000 | 0.009 | 0.386 | 0.000 | 0.977 | 0.947 |
| | Out | 0.073 | 0.033 | 0.033 | 0.073 | 0.345 | 0.566 |
| | precision(j) | 0.632 | 0.785 | 0.917 | 0.929 | | |

Figure A.6.: Screenshot of the online performance evaluator

In the confusion matrix, the red numbers indicate the incorrect or ambiguous estimations whereas the green numbers (in the diagonal of the matrix) indicate the accurate estimations. Concerning the derived criteria (*accuracy, precision, recall, gmean*), they are comprised between 0 and 1, for a quick outlook of their value, the higher they are, the greener they are colored and the lower they are, the redder they are colored.

To perform the simulation of the Location Tracking and its evaluation (as proposed in Section 4.3), the three processes are running in parallel: the Smart Home emulator, the Location Tracking simulator and the performance evaluator. Each process gives a graphical result in a separate window on the screen. The global result is composed of three figures: Fig. A.2 representing the Smart Home and the position, motion and action of the different inhabitants (in this case, there is one inhabitant in the bathroom), Fig. A.5 showing the estimated location, i.e. the active state(s) of the model (in this case, the location is accurately estimated and there is only one active state: $C$) and Fig. A.6 representing the results of the performance evaluator through the confusion matrix and the calculated criteria. The content of the three windows is updated online according to the inputs given by the user. When the user decided to close the program, the results are saved in a log-file.

## IV. Illustration on a simulated scenario

Based on this program, some scenarios can be tested on the considered case study and several instrumentations can be compared based on the evaluation criteria. As an example, a scenario involving a single inhabitant inside the house has been simulated. The Location Tracking has been performed using the model of Fig. A.5. The evolution of the different criteria along this scenario are given in Fig. A.7 (a) and Fig. A.7 (b) for the same duration of simulation.

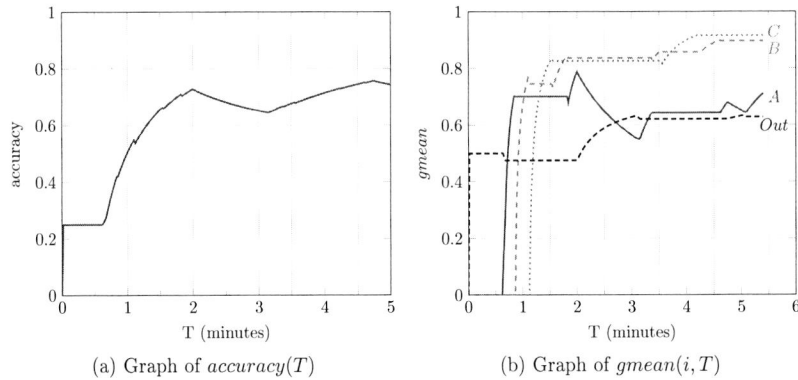

(a) Graph of $accuracy(T)$       (b) Graph of $gmean(i, T)$

Figure A.7.: Evolution of the overall $accuracy$ and of the local performances using $gmean(i, T)$ along the time $T$

The evolution of the $accuracy$ of Location Tracking (Fig. A.7 (a)) shows a quite good result at the end of the scenario. However, it can be seen that the $accuracy$ decreased between 2 and 3 minutes. The inhabitant has been simulated to go outside the house during this temporal window. Since there is no sensor outside in this case study, it is quite normal to have a decrease of the $accuracy$.

The evolution of $gmean(i, T)$ (Fig. A.7 (b)) confirms the previous result. The criterion $gmean$ shows great performances for the zones $B$ and $C$ (i.e. the bedroom and the bathroom) but worse performances for the zones $Out$ and $A$ (i.e. outside and in the living room). These results reflect a lack of instrumentation in these two zones or a bad placement of sensors. In this case, there is no sensor observing exclusively outside (lack of instrumentation) and the sensor ($DB$) is observing two zones ($A$ and $Out$).

These results provide the designer with indications to modify the zone partition and/or the instrumentation. For instance, it is possible to add a floor pressure sensor outside, just in front of the entrance door. By doing this modification and computing systematically the new models (not shown here), exactly the same scenario can be simulated using the new instrumentation and the new models. The results are shown in Fig. A.8 (a) for the *accuracy* and in Fig.A.8 (b) for the *gmean*.

Note that the *accuracy* at the end of this scenario has increased from 75% to 90% (Fig. A.8 (a)) and the different zones have an increased *gmean*, particularly the zones $A$ and $Out$ (Fig. A.8 (b)).

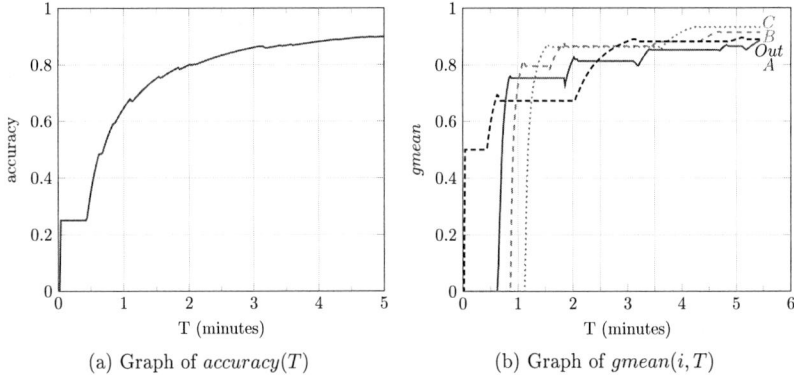

(a) Graph of $accuracy(T)$        (b) Graph of $gmean(i, T)$

Figure A.8.: Evolution of the overall *accuracy* and of the local performances using $gmean(i, T)$ along the time $T$ for the new instrumentation

The results provided by this dynamic evaluation are complementary to those given by the analytical approach. Considering the original and the modified instrumentation (with the additional floor pressure sensor), the analytical approach gives the same result, i.e. no *unlocationable zones* and Weak 1-Accurate-Location-Ability. However, the simulation-based criteria *accuracy* and *gmean* allowed highlighting differences between these two possible instrumentations and to quantify how good the new instrumentation is, compared to the original one.

## Conclusion

To conclude this appendix on software implementation, a whole software allowing performing simulation-based evaluation of Smart Homes for Location Tracking has been developed. It is composed of three processes running in parallel. A Smart Home emulator, allowing the user to immerse himself in a virtual Smart Home and allowing generating a sequence of sensor events corresponding to the desired scenario. A location tracker estimating the location of the inhabitants in real time, based on the sequence of sensor events. A performance evaluator, comparing the real location and the estimated location of the inhabitants.

As an outlook for future work, the library of sensors could be extended to not yet considered sensors. For now, only door barrier sensors, motion detectors and floor pressure sensors are considered and implemented. Furthermore, the model of functioning of the sensors could be improved in order to take into account sensor faults, for instance by integrating probabilistic models. Another improvement could be to integrate actuators in the Smart Home (automatic door, lights). Finally, the look of the Smart Home emulator and its realism could be improved, maybe by implementing a new emulator in 3 dimensions, using a more advanced videogame engine.

# Mickaël Danancher

# Lebenslauf

## Persönliches
Geboren am 7. Juli 1987 in Viriat (Frankreich)

## Schulbildung
| | |
|---|---|
| 05.07.2005 | Abitur (Baccalauréat, auf Französisch) |
| 09.2002 – 07.2005 | Lycée la Plaine de l'Ain, Ambérieu en Bugey (Frankreich) |

## Studium
| | |
|---|---|
| 31.06.2010 | Master of Science, Complex Systems Engineering |
| 09.2007 – 06.2010 | Studium des Maschinenbaus und der Automatisierungstechnik an der Ecole Normale Supérieure de Cachan (Frankreich) |
| 02.2010 – 07.2010 | Masterarbeit am Lehrstuhl für Automatisierungstechnik der TU Kaiserslautern (Deutschland) und am Automated Production Research Laboratory (LURPA) der Ecole Normale Supérieure de Cachan (Frankreich) |
| 04.2009 – 07.2009 | Studienarbeit am Lehrstuhl für Automatisierungstechnik der TU Kaiserslautern (Deutschland) |
| 09.2005 – 06.2007 | Vorbereitungsklasse, Physik und Technik, am Lycée la Martinière Monplaisir, Lyon (Frankreich) |

## Berufserfahrung
| | |
|---|---|
| Seit 09.2010 | Wissenschaftlicher Mitarbeiter am Automated Production Research Laboratory der Ecole Normale Supérieure de Cachan (Frankreich) und am Lehrstuhl für Automatisierungstechnik der TU Kaiserslautern (Deutschland) |
| 07.2005 – 08.2005 | Praktikant bei SECCA (Firma für Heizung und Klimatechnik), Ceyzériat (Frankreich) |

Cachan, December 2013